Optimization Aided Design

Optimization Aided Design

Reinforced Concrete

Georgios Gaganelis, Peter Mark, and Patrick Forman

Ernst & Sohn
A Wiley Brand

Authors

Dr.-Ing. Georgios Gaganelis
Univ.-Prof. Dr.-Ing. habil. Peter Mark
Dr.-Ing. Patrick Forman
Institute of Concrete Structures
Faculty of Civil and Environmental
Engineering
Ruhr University Bochum
Universitätsstraße 150
44780 Bochum
Germany

Coverfoto: Talbrücke Lindenau A44
Copyright: Aljona Riefert, KINKEL +
PARTNER, Dreieich, Germany
Photo editing: Patrick Forman, Razan
Karadaghi
Executing companies:
KINKEL + PARTNER, Dreieich,
Germany (structural design)
Züblin Hoch- und Brückenbau GmbH,
Bad Hersfeld, Germany (construction)

Unless otherwise stated, the rights to the
figures are held by the authors.

■ All books published by **Ernst & Sohn** are
carefully produced. Nevertheless, authors,
editors, and publisher do not warrant the
information contained in these books,
including this book, to be free of errors. Readers
are advised to keep in mind that statements,
data, illustrations, procedural details or other
items may inadvertently be inaccurate.

Library of Congress Card No.: applied for

British Library Cataloguing-in-Publication Data
A catalogue record for this book is available
from the British Library.

Bibliographic information published by
the Deutsche Nationalbibliothek
The Deutsche Nationalbibliothek lists
this publication in the Deutsche
Nationalbibliografie; detailed bibliographic
data are available on the Internet at
<http://dnb.d-nb.de>.

© 2022 Wilhelm Ernst & Sohn, Verlag für
Architektur und technische Wissenschaften
GmbH & Co. KG, Rotherstraße 21, 10245
Berlin, Germany

Print ISBN: 978-3-433-03337-1
ePDF ISBN: 978-3-433-61071-8
ePub ISBN: 978-3-433-61070-1
oBook ISBN: 978-3-433-61072-5

Typesetting   Straive, Chennai, India
Printing and Binding   CPI Group (UK) Ltd,
Croydon, CR0 4YY

Printed on acid-free paper

C9783433033371_100122

Foreword by Manfred Curbach

Intensifying Creativity in Construction

There is hardly a topic among building professionals that is discussed more intensively than sustainable construction. In view of the emphasis on this topic, it appears that intensive work is being done on the implementation of this challenge, both in research and in realization. After all, it is about nothing less than building in a way that enables all people of the generations to come to live a decent life on this earth. Because we have only this one. In 1994, the astronomer and astrophysicist Carl Sagan had the idea of taking a photo of the Earth with the help of the Voyager 1 space probe after it left the solar system. In a lecture on 13 October 1994 at the Cornell University, he said the following about this:

> Our planet is a lonely speck in the great enveloping cosmic dark. In our obscurity, in all this vastness, there is no hint that help will come from else-where to save us from ourselves. There is perhaps no better demonstration of the folly human conceits than this distant image of our tiny world. To me, it underscores our responsibility to deal more kindly with one another, and to preserve and cherish the pale blue dot, the only home we've ever known.[1]

In fact, we are overexploiting and consuming the resources of our earth and changing them massively. The consequences are climate change, scarcity of resources, natural disasters, hunger, flight, and misery. And the construction industry is massively involved in these developments.

The building industry in Germany accounted for 5.3% of nominal gross value added in 2018 (€179.6 billion GDP of €3388.2 billion GDP)[2] but causes around 25% of CO_2 emissions and uses around 40% of the energy generated.[3]

This discrepancy alone should lead to enormous productive activities. But what is the reality in terms of efficiency and research?

1 Sagan, C. (1997). *Pale Blue Dot*. United States: Random House USA Inc.
2 Statistisches Bundesamt (2019). *Bruttoinlandsprodukt 2018 für Deutschland*. Wiesbaden: Statistisches Bundesamt.
3 Hong, J., Shen, G.Q., Feng, Y., et al. (2015). Greenhouse gas emissions during the construction phase of a building: a case study in China. *Journal of Cleaner Production* 103: 249–259.

In sectors such as manufacturing (excluding construction), productivity increased by around 70% from 1995 to 2016, whereas in construction it only increased by around 5%.[4]

In terms of industry investment in research and development, out of a total of 436 571 people (in full-time equivalents) in 2017, only 1147 people came from the construction industry, i.e. 0.26%.[5]

The Federal Government of the Federal Republic of Germany spent a total of €17 250 million on research and development in 2018. Of this, €118.1 million was allocated to the area of "Regional planning and urban development; construction research," i.e. 0.69%.[6]

Considering only the Federal Ministry of Education and Research, a total of €10 486.7 million was invested in 2018. The area of "Regional planning and urban development; construction research" accounted for a share of only €27.5 million, i.e. 0.26%.

In 2019, the annual grant total from the German Research Foundation amounted to around €3285.3 million. The field of Civil Engineering and Architecture received grants totaling €51.5 million, i.e. 1.57%.[7]

The result of this small survey illustrates that in one of the most important industries in Germany, which contributes disproportionately to climate change, efficiency is stagnating and, at the same time, research is receiving severely below-average funding.

Every 12 years, the population of the earth grows by 1 billion people[8] who need a decent home, infrastructure, and energy supply. In view of the continuing increase in the world's population, we will not build less, but more. Contrary to this, we need to radically limit resource consumption and CO_2 emissions. It is obvious that in the future, building will have to be completely different, not just marginally, but fundamentally.

It is thus clear that we must significantly intensify research in the construction industry. Because of its enormous leverage effect, this is therefore one of the most important tasks for the future, both nationally and internationally, with extremely great significance for society as a whole. At all levels, from basic research to realization, for all available and newly to be developed building materials and combinations of building materials, in all areas of our social life up to politics, we have to become much more creative. Only through our inventiveness, our power of imagination for realization, our abilities to mentally penetrate complex processes

---

4 Statistisches Bundesamt (2020). Inlandsproduktberechnung – Lange Reihen ab 1970, Fachserie 18 Reihe 1.5. https://www.destatis.de/DE/Themen/Wirtschaft/Volkswirtschaftliche-Gesamtrechnungen-Inlandsprodukt/Publikationen/Downloads-Inlandsprodukt/inlandsprodukt-lange-reihen-pdf-2180150.html (accessed 27 August 2021).
5 Stifterverband für die Deutsche Wissenschaft e.V. (2019). *Forschung und Entwicklung in der Wirtschaft 2017*. Essen: SV Wissenschaftsstatistik GmbH.
6 Bundesministerium für Bildung und Forschung (2020) Bildung und Forschung in Zahlen 2020 (It should be noted that the figures therein for 2019 and 2020 are target figures).
7 Köster, T.; Lüers, K.; Schneeweiß, U.; Hohlfeld, C.; Kaufmann-Mainz, N. (2020). Deutsche Forschungsgemeinschaft - Jahresbericht 2019 – Aufgaben und Ergebnisse.
8 United Nations (2019). World Population Prospects: The 2019 Revision. https://www.dsw.org/infografiken/ (03 April 2021).

will we change the entire building process from design, planning, calculation, structure, material extraction, production, transport, on-site construction, operation, maintenance, data storage, strengthening up to further use, reuse, and recycling in such a way that we achieve climate- and resource-neutral building.

The methods, procedures, and calculations described in this book represent an important step toward a kind of building that has little to do with the way we know it today. And this is a good thing.

At the same time, may this book also promote the idea that it is worthwhile for everyone to think about change in the building industry to contribute ideas, to conduct research, and to work on realization. May the amount of research increase to a degree that is both appropriate and necessary to the challenge we all face.

Dresden, June 2021   *Manfred Curbach*

Prof. Dr.-Ing. Dr.-Ing. E.h. Manfred Curbach. Since 1994 professor for concrete structures at the Technische Universität Dresden, from 1999 to 2011 speaker of the SFB 528 "Textile Reinforcement for Structural Strengthening and Repair," since 2013 speaker of the BMBF consortium C^3 – Carbon Concrete Composite, since 2020 speaker of the SFB/TRR 280 "Design Strategies for Material-Minimised Carbon Reinforced Concrete Structures," 2014 foundation of the company CarboCon for the practical implementation of carbon concrete, 2016 winner of the German Future Prize, Award of the Federal President for Technology and Innovation.

Foreword by Werner Sobek

Building Emission-Free for More People with Less Material

Concrete is the only building material that can be cast into almost any shape on the construction site. It is available worldwide, is cheap, is easy to use, has a comparatively high strength, and is resistant to most environmental conditions. Concrete is the building material for everyone and for everything; it is the most widely used building material in the world. On the other hand, concrete is more unpopular for most people than almost any other material. In the past decades, this dislike was mainly based on its color, the quality of its surface, and its "coldness" (i.e. its low heat radiation). Today, it is the massive criticism of the CO_2 emissions caused by the production of cement, which, at around 8% of global CO_2 emissions, make a significant contribution to global warming and which, in terms of volume, even exceed the emissions of the entire global air traffic that labels concrete as an unloved, even demonized building material.

The currently widespread message that it will be possible to replace concrete as a building material with timber in the short to medium term is mostly based on an ignorance of the interrelationships. Mankind currently needs approximately 60–100 Gt of building materials per year in order to create a home for the new inhabitants on earth and to expand the existing built environment. This number does not include the so-called pent-up demand of the Third World, which with 6.3 billion people represents approx. 80% of the world's population and which, with a volume of approx. 60 t per capita, has a significantly lower building standard than the citizens of the industrialized nations, who account for approx. 335 t of building materials per capita.

If all the forests in the world were managed according to the principles of sustainable forestry, as is the case in many countries in Central Europe, for example, then a maximum of 10 Gt of construction timber could be obtained per year. An increased supply of construction timber through a higher logging rate in the forests would mean a reduction of the urgently needed CO_2 sink potential of these forests and must therefore be rejected. A redistribution of the available timber toward the construction industry would mean a reduction in the availability of wood to produce cellulose, paper or, for example, the abandonment of the cooking of daily food, as is still the case for many hundreds of millions of people every day.

To the scenario described above, the already mentioned pent-up demand for the inhabitants of the Third World. If the frequently voiced demand for an increase in prosperity and thus also a reduction in the birth rate in these countries were actually to be met through easier access to health care and education, especially for the female part of the population, the corresponding construction activities would have to be carried out, for example the building of schools and universities, medical practices and hospitals, including the associated infrastructure. Raising the level of construction in the Third World countries to that of today's industrialized countries can be estimated with a demand for building materials of 1700 Gt. This amount of building material represents twice the world built today. The climate-damaging emissions associated with the production of these building materials would make the earth uninhabitable for mankind. It is therefore evident that we will not be able to raise the total population of this earth to the building level of today's industrialized countries nor will timber as a building material be able to play a significant role in this context in the short to medium term. Timber will be an important building component in some parts of the world, especially in the Northern Hemisphere, in the short to medium term. Not more. Other building materials, such as clay or natural stone, will also increasingly find their way into construction. However, none of these materials will be able to replace concrete as a building material.

But what should the builders, the architects, the engineers, and the executing companies do, on the one hand, to fulfill their responsibility to provide a built home, including all the necessary infrastructural construction measures, for more and more people and, on the other hand, to make their enormously important contribution to limiting, even reducing, global warming? Since an ideal way has not yet been identified and the marvel material that solves all problems has not yet been found, the solution to the overall problem will consist of a sum of components. One of these components is the restriction of construction activities to what is actually necessary, the appropriate amount. This is often referred to as the principle of sufficiency. The principle of sufficiency includes the requirement not to demolish buildings or parts of them and replace them with new buildings until this is really unavoidable.

Another component of sustainable construction is the revolutionization of construction technology to the effect that in the future only recycling-oriented planning and construction will be permitted. In this way, the extraction of new building materials from the upper layers of the earth can be increasingly reduced in the medium to long term. This will also diminish, if not solve, the availability problems of individual building materials. It is common knowledge that enormous quantities of sand and gravel are required, especially for the production of concrete, and that sand has already become an extremely rare resource in some regions of our planet. The same applies to gravel and crushed stone. Immense availability problems are also expected for tin, zinc, and copper. For the construction industry, being the largest consumer of resources of all, it is therefore a matter of dramatically reducing the "consumption" of primary materials in the future and of using secondary materials where they are actually unavoidably needed. If we consider the availability of resources in addition to the emissions, it can be seen that the local and regional production of secondary material is associated with significantly lower climate-damaging emissions compared to primary material that is often delivered over long transport routes.

While the implementation of the closed-loop principle reduces the amount of "consumed" primary building materials, the complementary implementation of lightweight construction technologies can reduce the amount of consumed material and the amount of climate-damaging emissions during its production and distribution. This is where this book comes in. The introduction of state-of-the-art optimization methods and the resulting minimum-material component shapes, which also have a minimized need for reinforcing steel due to optimized reinforcement design, promote construction with concrete that is characterized by considerable material savings and thus considerable emission savings for the same utility value and durability. Supported by clearly understandable descriptions and a large number of examples, readers will find their way around quickly and easily. This makes it much easier to understand the subject matter, which is not always simple.

This book provides a significant contribution to establishing a new foundation for building with concrete, this wonderful building material for everyone and for almost everything. This foundation is characterized by the application of highly developed calculation methods and technologies that lead to material-minimized components and thus also to emission-minimized components. Both will be an essential part of tomorrow's construction, a construction that, like other sectors such as transportation or energy, must reduce its emissions by more than 50% by 2030. No one knows today how this will ultimately be achieved. However, the paths outlined in this book represent a valuable and indispensable tool on the way to achieving these goals.

Stuttgart, June 2021   *Werner Sobek*

Prof. em. Dr. Dr. E.h. Dr. h.c. Werner Sobek. From 1995 to 2021 successor of Frei Otto, and at the same time, from 2000 to 2021 also successor of Jörg Schlaich at the University of Stuttgart. Founder of the ILEK Institute of lightweight structures and conceptual design at the University of Stuttgart. From 2008 to 2014 professor at the Illinois Institute of Technology in Chicago in the succession of Ludwig Mies van der Rohe. Founder of the Werner Sobek group. Founder and co-founder of several charitable foundations and non-profit associations in the field of construction.

Preface

This book is based on over 15 years of research work at the Institute of Concrete Structures (Ruhr University Bochum, Germany) on topics related to structural optimization and lightweight concrete structures. The motivation, then and now, derives from two fundamental reasons. First, the climate challenge and the related necessity for lower material consumption. Second, modernizing the construction industry through new technologies aiming at more sustainable design and construction methods. The concepts evolved from the research work are combined in this book into an enhanced design approach, which we call *Optimization Aided Design* (OAD).

From students to researchers and practitioners, this book addresses everyone involved in structural engineering. Although the concepts primarily focus on concrete structures, they are generally adaptable to a wide range of further applications regardless of the material used. Numerous computational examples serve for a better comprehension of the methods and invite to discover the potential of OAD. Applications that have been successfully implemented further demonstrate transferability in practice and intend to provide inspiration for future projects.

Apart from the introduction, the book consists of two parts. The first part serves as introduction to the fundamentals of reinforced concrete design, on the one hand, and structural optimization on the other. Chapters 2 and 3 provide a general basis for understanding the methods presented subsequently. In no case do they claim to be exhaustive. For a more in-depth study of both topics, many excellent books from other colleagues already exist. In this regard, reference is made to the bibliography. The second part of the book introduces OAD for concrete structures. The methods are presented structurally from the outside in. In doing so, first, approaches for identifying the external structural shape are presented, followed by methods for designing the inner one (reinforcement layout), and finally techniques for the optimization of cross-sections.

Each of the OAD chapters is divided into three parts. They begin with a brief topical description supported by a representative overview figure, allowing the reader to decide whether the subsequent content has relevance for her or him. This is followed by the main section, in which the methods are discussed exhaustively and are supplemented with recommendations for their practical application. Numerous computation examples, to which reference is made in the respective main sections, provide the conclusion. They are further enhanced by application examples which have already been realized, for example ultra-light beams, extremely thin shells of solar thermal power plants or optimized reinforcement layouts for segmental tunnel linings.

OAD offers the possibility to enhance the daily engineering work and increase its efficiency. Our ambition is to highlight the great potential of the approach and thereby contribute to a modern, sustainable, and transparent way of designing and dimensioning reinforced concrete structures in the future. However, this can only succeed if we open the door for modern approaches and thus prove to the new generation, that the construction industry is able to adapt to the modern age. Considering the global challenges, let us be part of the solution, not part of the problem.

Bochum, Germany
April 2021

Georgios Gaganelis
Peter Mark
Patrick Forman

Contents

List of Examples

Acronyms

BS	block size
CONLIN	convex linearization
CTO	continuum topology optimization
CTTO	continuum truss topology optimization
FE	finite element
FEM	finite element method
FRC	fiber-reinforced concrete
GHG	greenhouse gas
HPC	high performance concrete
KKT	Karush-Kuhn-Tucker
MMA	method of moving asymptotes
MRM	material-replacement method
NC	normal strength concrete
OC	optimality criterion/criteria/condition
RC	reinforced concrete
SIMP	solid isotropic material with penalization
SLP	sequential linear programming
SLS	serviceability limit state
SQP	sequential quadratic programming
STMs	strut-and-tie model
TTO	truss topology optimization
UHPC	ultra-high performance concrete
UHPFRC	ultra-high performance fiber-reinforced concrete
ULS	ultimate limit state

About the Authors

Georgios Gaganelis is a structural designer for civil engineering structures and a freelance consultant in structural optimization. In 2020, he received his PhD at the Ruhr University Bochum, Germany in the field of optimization strategies for concrete and steel–concrete composite structures. His research interest focuses on topology optimization and material-driven steering. A special focus lies on ultra-light structures requiring minimal material efforts.

Peter Mark is a full professor for Structural Concrete at the Ruhr University Bochum, Germany. He is researching on applied optimization methods and lightweight concrete structures since 20 years. He received his PhD in 1994 and the postdoctoral degree in 2006. He is Consultant Engineer and Independent Checking Engineer since 2008 and involved in several bridge, tunnel, and building construction projects.

Patrick Forman is a postdoctoral research fellow at the Institute of Concrete Structures at Ruhr University Bochum, Germany. He received his PhD in 2016. More than 10 years, he is researching on lightweight shell and beam structures made of high performance materials using various structural optimization techniques. Currently, he is technical and managing director of an interdisciplinary research center on adaptive modularized construction methods.

Acknowledgments

This book comprises the research results of numerous completed and ongoing doctoral theses at the Institute of Concrete Structures, Ruhr University Bochum, Germany. We would like to thank all research assistants whose work has contributed to the concept of optimization aided design (OAD) for concrete structures, namely Dr.-Ing. T. Putke, C. Kämper, M.Sc., Dr.-Ing. M. Smarslik, D. Petraroia, M.Sc., as well as the many student assistants who participated in the projects.

The German Research Foundation (DFG) is thanked for its financial support of the research projects. In particular, these are the Priority Programme 1542 "Concrete Light" (SPP 1542), the associated transfer projects "Light, microreinforced UHPC shell structures optimised to deformations using the example of parabolic trough collectors in concentrating solar power plants" and "Pinpoint accurate adaptive structures for heliostats made from high performance concrete for central tower power plants," as well as the Collaborative Research Center 837 "Interaction Modeling in Mechanized Tunneling". The authors would also like to thank the German Federal Ministry of Economic Affairs and Energy (BMWi) for the financial support of the project "ConSol – Concrete Solar Collector" in the framework of the 6th Energy Research Programme on the basis of a decision by the German Bundestag.

Sincere thanks are further given to all project partners and sponsors, especially Prof. Dr.-Ing. Dr.-Ing. E.h. J. Schnell, Dr.-Ing. S. Penkert, Dr.-Ing. R. Buck, Dr.-Ing. Dipl.-Theol. A. Pfahl, Dr.-Ing. habil. B. Sagmeister, Dipl.-Ing. T. Friedrich, Dr. R. Dasbach, Dipl.-Ing. T. Deuse, Dipl.-Ing. K. Hennecke, Dipl.-Ing. D. Krüger, T. Zippler, M.Sc., Dipl.-Ing. M. Kintscher, Dipl.-Ing. A. Natterer, and Dipl.-Ing. T. Stanecker as well as the institutions and companies TU Kaiserslautern, German Aerospace Center (DLR), Almeco, Pfeifer Seil- und Hebetechnik, Solarlite CSP Technology, Stanecker Betonfertigteilwerk, durcrete, Innogration, and Dyckerhoff.

We would also like to thank the Ernst & Sohn publishing house for making this book project possible. In particular, we thank F. Stürmer, Dipl.-Ing. C. Ozimek and U.-M. Günther for the pleasant cooperation.

Finally, we would like to express our particular gratitude to Prof. Dr.-Ing. Dr.-Ing. E.h. M. Curbach and Prof. Dr. Dr. E.h. Dr. h.c. Werner Sobek, who collaboratively wrote the foreword to this book. We are exceptionally honored to include the thoughts of two luminaries of civil engineering and thought leaders on sustainability in structural engineering in this book.

Georgios Gaganelis
Peter Mark
Patrick Forman

1

Introduction

Key learnings after reading this chapter:

> - What influence do construction activities exert on the climate?
> - What are the requirements of a future-oriented way of building?
> - How can structural optimization methods contribute to this?

1.1   Preliminaries

Concrete is one of the oldest building materials in the world. The Romans already used it and achieved strengths of up to about 40 MPa [1]. Concrete exhibits an uneven distribution of strengths. Its compressive strength is high, but its tensile strength equals about one tenth of it. Concrete is therefore predominantly stressed in compression while tension is avoided. Structures whose design is oriented toward the material properties take this aspect into account and primarily bear loads via compressive stresses. Thus, this corresponds to the traditional use of concrete for arches, columns, and domes, where compressive forces prevail.

With the advent of reinforced concrete (RC) in the 19th century, the restriction to structures subjected predominantly to compressive stress was lifted [2], since steel was employed to enhance concrete in tension emerging from bending, shear forces, or axial loads. The same applies to local areas of load application or geometric discontinuities such as corbels, openings, and supports.

Due to three factors, namely the technological feasibility of reinforcement, scarcity of human labor and low material costs for concrete itself, the structures changed. Curved shapes disappeared and were widely substituted by simple, rectangular ones, which prevail to the present day. Typical examples are prismatic plates, walls, or beams. The formwork shapes are simplistic and can be easily produced on a large scale. Thus, the structural designs are often motivated by the simpler construction on site, although larger quantities of concrete and steel are needed.

Building is a question of time, its capabilities and the pursuit of goals [3]. Expensive materials lead to rather slender designs. High labor and formwork costs require automation and simplification of shapes. Furthermore, the demand for short construction times favors prefabricated components.

Optimization Aided Design: Reinforced Concrete, First Edition.
Georgios Gaganelis, Peter Mark, and Patrick Forman.
© 2022 Ernst & Sohn GmbH & Co.KG. Published 2022 by Ernst & Sohn GmbH & Co.KG.

Compared to other building materials, concrete has the advantage of free formability. The initially liquid material hardens and obtains its permanent, solid shape
only through hydration. In this respect, concrete structures are open to virtually
unlimited outer shapes. In the same way, its inner shape – namely the layout of
reinforcement – is freely adjustable and installable. It does not have to adapt to
formwork edges in any way, but can easily follow the tensile trajectories instead.

So what is the right shape? What is the right reinforcement (pattern) and where
is it located? How does the right cross-section look like? The answers to these
questions depend on the defined objectives and constraints to be met. And these
conditions change. They change over time, they change depending on the country
and the way of life, and they change according to external influences. Because of
this mutability, the approaches to find the right shapes must therefore account
for the relevant objectives and constraints. This is the only way to find different
answers to similar questions, differing only in the underlying boundary conditions.
In this way, the range of answers expands and adapts to the scope.

The book develops a generally applicable method for this purpose, namely the
Optimization Aided Design (OAD). It is motivated by three fields of investigation.
These are:

- the outer shape of structures,
- the inner reinforcement layout, and
- the cross-sectional design.

Briefly, the inner and outer form finding.

1.2 Outer and Inner Shaping

The outer shape of a concrete structure is determined by its intended purpose. This
can be, for example, the enclosure of a storeroom, the load-bearing function for a
bridge deck, a plane surface of a ceiling, or shielding against soil and groundwater
in a tunnel. The shape further characterizes the design.

The outer shape is influenced by the load-bearing and protective function to
be met, aspects of economic efficiency, the construction process and the way the
structure integrates into the surrounding environment. It is virtually arbitrary due
to the ability of free formability, since concrete exhibits the advantage of taking any
shape. In this respect, no distinction can be made between "right" and "wrong",
since all designs meet the requirements for load-bearing capacity, serviceability,
and durability.

In the case of engineering structures such as bridges or tunnels, their outer
shape clearly indicates the static load-bearing system. If the outer form follows the
dominant internal force flow ("form follows force") and further predominantly
orients toward compression, lightweight and, in many cases, bionic-like, curved
structures result. Typical examples are arches, shells or even pylons of bridges with
inclined cables, which experience a distinctly dominant downward load impact due
to the deflecting forces of the cables.

Figure 1.1 Concrete shaped to the flow of forces: pylon of the Pont de Terenez (France). Picture: Thomas Putke-Hohmann.

Figure 1.1 shows such a form finding according to the flow of forces by the bridge "Pont de Terenz", located in France [4]. The bridge layout lies within a narrow circular bend. The pylon is shaped similar to an inverted "Y", with the front leg continuing the slightly inclined column shape above the roadway level in a straight line and orienting toward the lateral forces from the cable deviation. In this way, it is mainly subjected to compressive loads. The lower supports act like two legs in step sequence and provide further stability against lateral loads from wind and centrifugal downforces.

Structures designed according to the flow of forces tend to look aesthetically pleasing and require small amounts of material by avoiding redundant flexural load-bearing effects. In contrast, Figure 1.2 shows a solid bridge designed using simple shapes. It is composed of equal single-span beams, each of which is interchangeable. The external cross-sectional geometry of the superstructure remains constant. This results in two bearing rows per column that are accordingly thick and compact.

The bridge can be built with constant scaffolding and formwork layout. The dominant design aspect here is an effective fabrication with a repetitive sequence,

Figure 1.2 Railway concrete bridge with uniform single span girders and compact columns.

meaning that every single beam is made with the same formwork and scaffolding, the same reinforcement and the same casting concept. The result is a repetitive pattern justified from cost-efficiency in manufacturing. Due to the homogeneous shape across the entire bridge, its beam-like load-bearing behavior and the necessary limitation of distinct rotation angles at the ends of each girder for the train passage, massive, heavy cross-sections result and the spacings between the columns become compact and monotonous. Compared to a structure following the flow of forces, great additional material amount is needed. For the specific case, this is mass concrete, as well as prestressing and reinforcing steel. At the same time, aesthetics suffer due to the obstructed view and consistent pattern of equal spans, equal columns, and the tall and heavy-looking superstructure.

Similar to the external shape, the internal load transfer can also be designed intentionally. The reinforcement is typically adapted to the formwork shape and not primarily to the tensile trajectories of the load transfer. For common rectangular cross-sections, reinforcement cages are produced from longitudinal bars and vertical stirrups, thus forming a rectangular grid that follows the formwork's edges (Figure 1.3a). This type of reinforcement suits the convenience of production. Due to the constant orientation, tensile trajectories and reinforcement directions differ, which automatically leads to an increased amount of required reinforcement. In other words, fabrication outweighs material efficiency.

Figure 1.3b shows an alternative layout. Like the outer shape of the concrete cross-section, the reinforcement can also be laid in a virtually arbitrary spatial manner and does not necessarily has to follow a rectangular pattern. The depicted

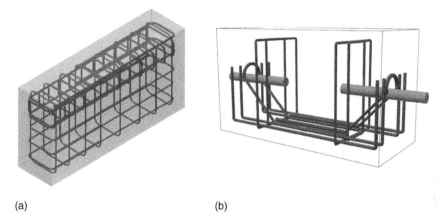

(a)   (b)

Figure 1.3   Reinforcement cages: (a) typical rectangular pattern of longitudinal bars and side aligned stirrups, (b) freely bend reinforcement derived from strut-and-tie modeling [5].

shear force transfer reinforcement layout with shear dowels was developed from a strut-and-tie model adapted to the force flow [6].

Generally, it can be stated that the outer and the inner shape of RC structures are not limited. If they follow the flow of forces, low material consumption and mostly aesthetic designs result.

The following example of a cross-section choice will illustrate how much the underlying objectives influence the design. A RC cross-section for a beam in axial bending is sought. The cross-section needs to be both sustainable and cost-effectively designed for various monetary boundary conditions, however, the bending moment remains equal in each case. Figure 1.4a shows a cross-section with expensive formwork and subordinate material prices for both concrete and steel reinforcement. The result is a simple rectangular shape. In Figure 1.4b, the high price for steel is the governing criterion. A more elaborate I-shape exhibiting a large lever arm of internal forces now minimizes the reinforcement amount. The third cross-section design (Figure 1.4c) arises from the scenario, where the price for concrete is dominant. Here, the web width decreases, the top flange is thinned,

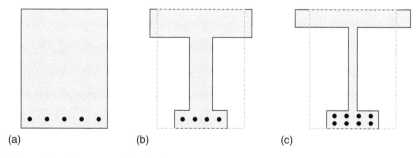

(a)   (b)   (c)

Figure 1.4   RC cross section designed under cost boundaries: (a) expensive formwork, (b) expensive reinforcing steel, (c) expensive concrete.

and further reinforcing steel is added in order to shift the neutral axis of elongation down into the web.

This simple example demonstrates the need to formulate distinct and measurable objectives in order to find adequate structural designs for different boundary conditions. Although, a fixed preliminary specification of the cross-section satisfies load-bearing capacity, serviceability and durability, it is not capable to respect other, often multifunctional requirements, concerning, for instance, material, fabrication, and time. For this purpose, load-bearing capacity, serviceability, and durability must be downgraded to become auxiliary conditions to the design problem. They must be met in any case but not as primary design objectives. Instead, the aim must be to find the right shape, freely according to descriptive objectives such as, e.g. cost minimization or short construction times.

1.3 Environmental Demands

Concrete is by far the most widely used building material in the world [7, 8]. Its major advantages are first, the easy availability of the raw materials, namely cement, aggregates, and water, second, the free formability, third, the simple handling and, fourth, most importantly, its low price. Even bottled water is more expensive per unit volume than concrete.

The amount of concrete used worldwide each year is gigantic, estimated at more than 10 billion tons in 2019 [9]. This equals a per capita consumption of about 1.3 t/a (tons per year) – 1 for each of the almost 8 billion people on earth. If the total annual amount were to be distributed over a land area such as that of Germany ($\sim$ 360 000 km^2), it would be covered with about 1.4 cm of concrete. Taking the more reliable data of the total amount of globally produced cement for the calculation (4.1 billion tons in 2017 [10]) and assuming a proportion of 300 kg of cement per cubic meter of concrete, this even leads to 3.8 cm of concrete cover. An unimaginable amount of material per year.

However, when it comes to its impact on the climate, concrete is not a particularly unfavorable material. In fact, rather the opposite is the case [11]. With usually short transport distances and almost worldwide availability of the raw materials, concrete proves to be very efficient compared to other building materials in terms of the

Table 1.1 Specific strengths of construction materials: normal strength concrete (NC), structural steel, high performance concrete (HPC), ultra-high performance concrete (UHPC), fiberglass, carbon.

	NC	Structural steel	HPC	UHPC	Fiberglass	Carbon
Strength [MPa]	16–50	235–355	55–100	150–250	1200–1300	2000–6000
Density [t/m³]	2.1–2.6	7.85	2.1–2.6	2.4–2.6	1.9	1.8
Specific strength [MNm/t]	6–24	30–45	21–48	58–104	632–684	1111–3333

load-bearing capacity provided per unit volume ("specific strength"), see Table 1.1. Concrete becomes a highly climate-relevant construction material only due to its vast quantity. The production of cement alone, its most important raw component, is responsible for 5–10% of the global CO_2 emissions each year [12–15]. The combustion process in cement production is thus the largest single emitter worldwide.

The increase in global use of concrete is closely linked to the beginning of industrialization in the 19th and 20th centuries and the associated population growth. Figure 1.5 shows the development of the main raw materials of concrete over the years. These are water, cement and crushed or round aggregates. Concrete consumption has increased significantly. Its application has been particularly intensified in the last three decades throughout the world. Closely related to concrete is the demand for aggregates, which account for around 70% of its volume. Over 40 billion tons are used each year, which equals about half of the total sand

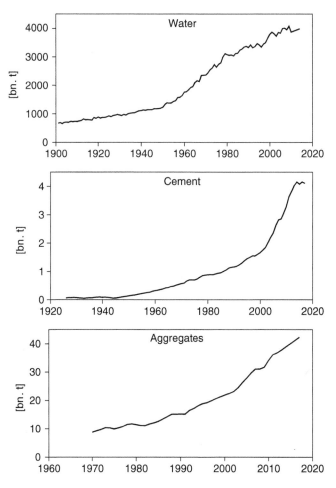

Figure 1.5 Annual world consumption of water, cement, and aggregates, according to [10, 16, 17].

and grain mining [17]. The consumption of cement and water has increased sharply within the last decades, which is closely related to concrete consumption. It should be noted, however, that Figure 1.5 depicts the total water use worldwide, meaning that it also includes the use for drinking water, irrigation, livestock, and other industries.

At the same time, global population is growing. Growth increased sharply after the two World Wars and has followed an almost linear progression since the 1960s. Figure 1.6 shows the development until 2020 with a mean prediction until 2100 according to estimations of the United Nations [18]. Following an average prognosis, about 1 billion people are added every 10–15 years. In 2100, approximately 3.2 billion more people are expected to live on earth than do today. Simultaneously, the living standard of people is increasing and along with that the need for housing, infrastructure, and energy supply. By comparison, the population increase from 2020 to 2050 will be around the total number of people in the world in 1930. At that time, about 2.0 billion people lived on earth. About the same volume of housing and associated infrastructure at that time is to be provided over the next 30 years in addition to the current state [19, 20]. However, this additional amount represents rather a lower estimate of demand, since rising living standards are not included in the calculation.

New construction is increasingly being accompanied by the necessary preservation of existing structures as an ongoing task. On the one hand, bearing structures – like all technical equipment – must be regularly maintained and repaired. On the other hand, they have a limited service life. The service life varies from about 50 years for buildings to 200 years for extraordinary infrastructure or utility constructions. After their use, structures need to be replaced by new ones.

The substitution of existing structures by new ones is called replacement construction. It is typical for structural tasks in, for example, Western Europe or North America. Replacement constructions are built at the same place as the structures to be replaced. Consequently, construction activities take place in a dense urban

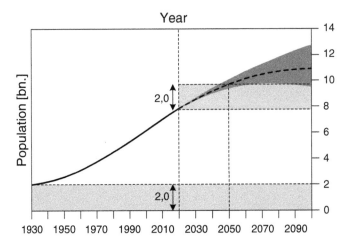

Figure 1.6 Development of the world population from 1930 till 2100, according to [18].

Figure 1.7 Traffic jam caused by a lane constriction to pass a construction site of a road bridge replacement.

environment and in existing road or utility networks. Inevitably, this leads to restrictions and disruptions in the flow of goods and traffic.

This is particularly evident at bridge construction sites. The ongoing traffic is to be diverted around the construction site and the lanes must be restricted. Figure 1.7 shows a typical example of a lane constriction at a construction site of a road bridge. Truck and passenger traffic is narrowed from initially three lanes to only one. This causes vehicles to back up a long way. The result is a waste of time, human labor, and fossil fuel – without generating any economic gain.

Replacement construction rarely takes place consistently. Consistency in this context refers to a distribution over time, across locations and in terms of monetary and personnel efforts. Usually, construction activities tend to take place in waves and are therefore clustered in places of particular economic growth and are focused on few years. This hardly makes sense for builders, contractors, and planners. Figure 1.8 shows an example of bridge building activities in a western German city. Gray columns represent amounts of installed bridge deck areas. Bridge construction is concentrated around the 1960s to 1980s and the distribution resembles a bell-shaped curve. If the bridges are replaced after a service life of 80 years, which is assumed to be the same for all structures in a simplified manner, the distribution of replacement construction shown on the right in light gray is obtained. It repeats the dark gray bars and merely shifts them 80 years into the future. The unfavorably clustered construction activities repeat. Now, however, they take place within the existing traffic network. They thus entail considerable restrictions from the

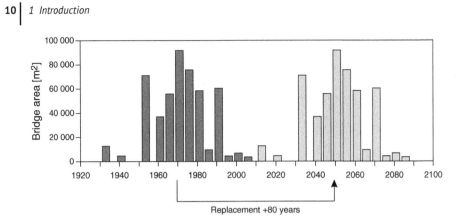

Figure 1.8 Age distribution of built bridges of a German city according to [21] and proportional prognosis of its rebuild after 80 years of service life.

significantly increased traffic over the recent years and the ever more densely linked flows of goods of industrial supply delivered "just in time". Extending the service life of structures to far beyond 100 years seems to offer a reasonable solution in order to distribute construction activities more evenly over time [22]. Generally, the key issue lies in a more uniform distribution. This results in almost constant demands on building, costs, and human engagement.

Building is further associated with a high personnel expenditure. Many activities are still conducted by hand and structures hardly involve serial repetition. As an example, Figure 1.9 shows several workers distributing and planning a concrete subbase for a tunnel trough. Human labor is becoming a rare commodity, particularly in industrialized countries, and must be conserved in the same way as

Figure 1.9 Workers manually distributing and planning a concrete subbase.

materials, time, money, and climate sensitivity. This endeavor will become even more important in the future than it has been in the past.

The examples of current construction serve to illustrate that objectives of reasonable structural designs may depend on various boundary conditions. This may be saving of materials, but also minimizing construction times, human labor costs, and greenhouse gas emissions or ensuring more uniformly distributed construction activities. Building reflects the time, its respective boundary conditions, as well as primary and secondary objectives. However, both objectives and constraints may change over time, but this does not apply to their fundamental components, namely objectives and constraints. For this reason, it seems reasonable to define the design and dimensioning of structures following a corresponding approach, namely by using objectives consisting of one or multiple (weighted) aims and boundary conditions to be met (constraints).

OAD provides such a uniform approach. One or more goals are formulated as the objective function. Typical examples include minimizing the total costs, weight, structural compliance, or manufacturing labor. In doing so, various boundary conditions must be complied with. Typical examples for the latter are stress limits, ensuring equilibrium, restricting the material amount, or geometric limits. This reveals a generally applicable method toward finding structural designs capable of meeting the complex requirements of the modern world more effectively.

1.4   Optimization Aided Design (OAD)

OAD refers to the design and dimensioning of structures and structural components using optimization methods. For RC, this involves finding the most suitable outer shapes, identifying efficient reinforcement layouts, and employing reasonable cross-sections. OAD therefore refers to a general methodology that can be applied to a great extent. It is suitable for the use with any concrete (normal strength to ultra-high performance) and arbitrary types of reinforcement (steel, carbon, fibers), and is further transferable to numerous fields of structural engineering such as steel, timber or foundation engineering.

OAD uses the concept of mathematical optimization [23–28]. In doing so, generally, the aim is to minimize the formulated objective function. This can be, for example, the structural weight, a reinforcement quantity, or even combinations of conflicting functions, for instance, minimizing material usage, while also minimizing shuttering edges. Additionally, several restrictions must be met. These restrictions take the form of equality and inequality constraints. Typical equality constraints are the equilibrium between external actions and internal forces. Inequality constraints may include geometric limits, stress constraints, or an upper limit of the material amount. Within mathematical optimization, OAD relies on structural optimization approaches, and focuses in particular on the topology optimization method.

As an introduction, Figure 1.10 shows the gradual adaptation of the outer shape of a single span girder under uniform loading. The designs are depicted along with

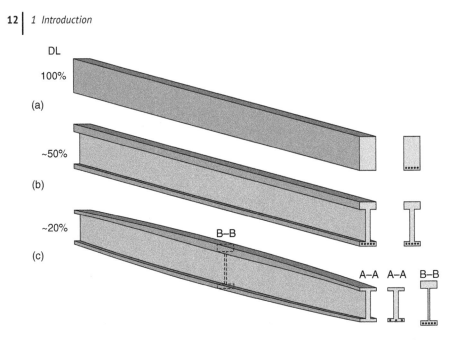

Figure 1.10 Stepwise shaping of a single span girder with reduction in dead loads: (a) constant rectangular cross section, (b) shaping of the web, (c) force affine shaping of heights, flanges and the web thickness.

their respective cross-sections and the relative change in dead loads (DL). In order to illustrate the concept, the example is intentionally kept simple and can in principle be solved intuitively without applying any optimization technique. Starting point is a simple beam with rectangular cross-section (a). Decisive for the bending design is the bending moment at midspan. The shear reinforcement is determined close to the supports from the maximum shear force. The outer shape of the structure is dictated by the respective required cross-section. In doing so, excess material is used almost throughout the entire beam since only two particular cross-sections are decisive. The web is generally too thick, the maximum flexural reinforcement is only required in the center of the field, as is the resultant in the compression zone, which decreases considerably toward the bearings. In the second step (b), the web is reduced to the actual necessary width from the load-bearing capacity of the inclined compression strut. Although, the beam remains prismatic in its outer shape, the DL is already decreasing considerably. A slight increase in height also maintains the stiffness. In the third step (c), each cross-section of the beam is formed in such a way that compression zone, tension zone, web width and stirrup quantity are always fully utilized. The shape in the longitudinal direction is adapted to the parabolic shape of the bending moment, the web width to the respective inclined compression force and the flanges to the necessary widths for compression force and tensile reinforcement. Single cross-sections are no longer decisive. Instead, each cross-section along the entire length of the beam is now relevant, which also means that reserves are no longer available. A reduction of the DL up to about 20% is achieved compared to the simple beam with rectangular cross-section.

Conceptually, OAD follows the same approach: searching for design solutions that ensure efficient stress utilization and accumulate material only where it is most effective for the load-bearing mechanism. This applies equally to the outer and inner form finding.

For typical structures, several load cases with varying actions have to be taken into account, thus full utilization in superposition of different scenarios, for instance, maximum vertical loads versus maximum horizontal loads, obviously may not be achieved everywhere.

Taking into account the relevant boundary conditions is particularly important in OAD. Such boundary conditions are especially geometric limits such as clearances and smooth edges for connection, but also maximum and minimum dimensions. In the case of a ceiling slab, for example, the flat, smooth surface is mandatory and curved shapes or recesses must be prevented, even if they would favor the load-bearing mechanism.

Figure 1.11 shows an example for the influence of the geometric boundary conditions of a roof girder. The static system with permissible design space (gray) and the imposed action consisting of several point loads F are depicted in (a). From the topological optimization, minimizing the mean structural compliance (=maximizing stiffness) and limiting the allowed material amount, leads to the density distribution shown in (b), which essentially represents a load-bearing structure consisting of axially loaded struts. Bending is thus decomposed into a truss-like system including tension and compression members. The density distribution serves as an inspiration for the design. This is interpreted into two different variants shown in Figure 1.11c. On the left, a mostly clear truss structure is shown, which follows the optimization result. Alternatively, a RC solution is developed on the right, which transfers the single compression and tension struts

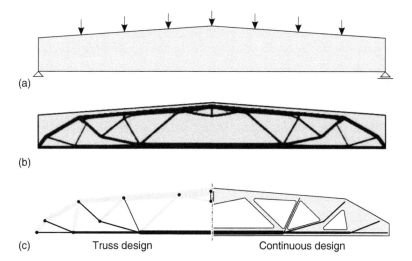

(a)

(b)

(c)   Truss design   Continuous design

Figure 1.11 OAD of a roof girder: (a) static system and design space, (b) optimization result, (c) resulting truss structure and RC structure with free reinforcement layout.

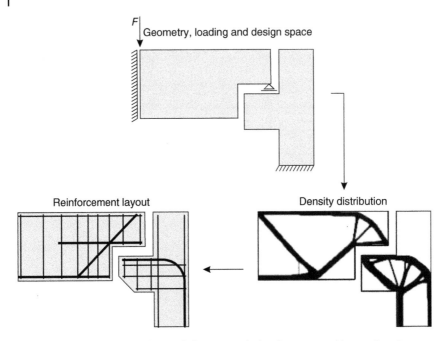

Figure 1.12 Optimization aided reinforcement design for a stepped beam placed on a corbel.

into concrete and reinforcement areas, respectively. This results in a structure with variable cross-sections, block-outs and a free reinforcement layout.

For the inner reinforcement design, OAD leads to concepts similar to those known and proven from strut-and-tie models.

Figure 1.12 shows the example of a stepped beam that is supported by the corbel of a column and is loaded in bending. The applied topology optimization yields a density distribution that exhibits partly truss-like and bionically curved shapes. Stress evaluation using the Finite Element Method (FEM) reveals tension and compression struts, representing concrete or reinforcement areas. From this, a reinforcement design can be derived for both cases. Alternatively, for implicit quantitative evaluation, more advanced truss topology optimization approaches are ideal. They offer the benefit of assigning a specific force to each member, which can then be directly converted into required reinforcement and concrete areas.

1.5 Structure of the Book

Apart from this introduction, the book is divided into five more chapters.

Chapters 2 and 3 briefly summarize the main principles of RC design and structural optimization. For Chapter 2 this involves the material behavior of concrete and reinforcement and basic RC design concepts. Chapter 3 deals with the basic optimization procedure, fundamental formulation of the optimization problem as well as important concepts like the Lagrange function, sensitivity analysis, and general solution methods.

The focus of Chapter 4 lies on the identification of the outer shape of structures. Various approaches are introduced for one and bi-material structures. Moreover, concepts for steering one-material structures toward a compression or tension dominant load-bearing behavior as well as bi-material designs including tension-only and compression-only materials like reinforcement and concrete, respectively, are introduced. The resulting material distributions yield, on the one hand, structural designs containing axially loaded members, like truss-like structures and arcs, and, on the other hand, apply the materials with respect to their particular stress-affinities. Multiple numerical examples illustrate the approaches. In addition, several applications demonstrate the applicability and great resource saving potential for structures in practice.

Chapter 5 introduces approaches seeking the inner shape. This essentially includes identifying the internal force flow and deriving suitable strut-and-tie models. In doing so, some of the methods already described in the previous chapter are adapted to the different scope. However, also an alternative approach based on truss topology optimization is presented, which holds several advantages over the continuum-based methods. In addition, especially for particular challenging problems, a more sophisticated topology optimization approach, employing both continuum and truss finite elements within optimization, is derived. Again, numerous examples and applications emphasize the applicability.

Chapter 6 deals with optimized design of cross-sections. The aim is to find the optimal shape and design while complying with stress and strain limits. The starting point is an arbitrary composite cross-section under axial or biaxial bending with and without normal force. In a second step, the concept is limited to RC cross-sections. It is shown how optimization methods can be applied to identify the right cross-sectional shape, determine reinforcement amounts and solve the equilibrium iterations in order to compute the strain distribution. The method is applied to several examples. They contain primarily RC cross-sections, but also foundations, in order to demonstrate the general applicability. The aim is to improve the shape in order to save material and implement particularly effective designs. The actual verification of the load-bearing capacity is provided automatically, since strain and stress limitations are met via boundary conditions.

References

1 Lamprecht, H.-O. (2001). *Opus caementitium: Bautechnik der Römer*. Düsseldorf: Bau + Technik. ISBN: 3764003502.

2 Mörsch, E. (1909). *Der Eisenbetonbau: Seine Theorie und Anwendung*. The Engineering News Publishing Company.

3 Gaganelis, G., Forman, P., and Mark, P. (2021). Stahlbeton optimiert - für ein Mehr an Weniger. In: *Nachhaltigkeit, Ressourceneffizienz und Klimaschutz*

(ed. B. Hauke, Institut Bauen und Umwelt e.V., and DGNB e.V.). Ernst & Sohn, 159–167. ISBN: 9783433033357.

4 Virlogeux, M., Motard, M., Bouchon, E. et al. (2004). Design of the Terenez cable-stayed bridge. In: Fédération internationale du béton (fib) *Concrete Structures*, Avignon, France: FIB/CEB-FIP, 228–229.

5 Putke, T. (2016). Optimierungsgestützter Entwurf von Stahlbetonbauteilen am Beispiel von Tunnelschalen. PhD thesis. Bochum: Ruhr University Bochum.

6 Putke, T., Bohun, R., and Mark, P. (2015). Experimental analyses of an optimized shear load transfer in the circumferential joints of concrete segmental linings. *Structural Concrete* 16 (4): 572–582. https://doi.org/10.1002/suco.201500013.

7 Curbach, M. (2013). Bauen für die Zukunft. *Beton- und Stahlbetonbau* 108 (11): 751. https://doi.org/10.1002/best.201390098.

8 World Business Council for Sustainable Development (2016). Cement Industry Energy and CO_2 Performance: Getting the Numbers Right. https://www.wbcsd.org/Sector-Projects/Cement-Sustainability-Initiative/Resources/Cement-Industry-Energy-and-CO2-Performance (accessed 25 August 2021).

9 Byatt, G. (2019) Improving the world through better use of concrete. https://irp-cdn.multiscreensite.com/8bbcaf75/files/uploaded/190805_MRS-1_concrete.pdf (accessed 25 August 2021).

10 Kelly, T.D. and Matos, G.R. (2016). Historical statistics for mineral and material commodities in the United States. https://www.usgs.gov/centers/nmic/historical-statistics-mineral-and-material-commodities-united-states (accessed 5 November 2021).

11 Hammond, G.P. and Jones, C.I. (2008). Embodied energy and carbon in construction materials. *Proceedings of the Institution of Civil Engineers - Energy* 161 (2): 87–98. https://doi.org/10.1680/ener.2008.161.2.87.

12 Mehta, P.K. (2001). Reducing the environmental impact of concrete. *Concrete International* 23 (10): 61–66.

13 Rodrigues, F.A. and Joékes, I. (2011). Cement industry: sustainability, challenges and perspectives. *Environmental Chemistry Letters* 9 (2): 151–166. https://doi.org/10.1007/s10311-010-0302-2.

14 Suhendro, B. (2014). Toward green concrete for better sustainable environment. *Procedia Engineering* 95: 305–320. https://doi.org/10.1016/j.proeng.2014.12.190.

15 Worrell, E., Price, L., Martin, N. et al. (2001). Carbon dioxide emissions from the global cement industry. *Annual Review of Energy and the Environment* 26 (1): 303–329. https://doi.org/10.1146/annurev.energy.26.1.303.

16 Ritchie, H. and Roser, M. (2017). Water Use and Stress. https://ourworldindata.org/water-use-stress (accessed 25 August 2021).

17 United Nations Environment Programme (2021). International Resource Panel: Global Material Flows Database. https://www.resourcepanel.org/global-material-flows-database (accessed 25 August 2021).

18 United Nations, Department of Economic and Social Affairs, Population Division (2019). *World Population Prospects 2019*. New York: United Nations. ISBN: 978-92-1-148316-1.

19 Sobek, W. (2015). Die Zukunft des Leichtbaus: Herausforderungen und mögliche Entwicklungen. *Bautechnik* 92 (12): 879–882.

20 Sobek, W. (2021). Bauen für die Welt von morgen. In: *Nachhaltigkeit, Ressourceneffizienz und Klimaschutz* (ed. B. Hauke, Institut Bauen und Umwelt e.V., and DGNB e.V.). Ernst & Sohn, 31–38. ISBN: 9783433033357.

21 Stratmann, R., Birtel, V., Mark, P. et al. (2008). A digital system to record and analyse cracks. *Beton- und Stahlbetonbau* 103 (4): 252–261.

22 Curbach, M., Bergmeister, K., and Mark, P. (2021). Baukulturingenieure. In: *Nachhaltigkeit, Ressourceneffizienz und Klimaschutz* (ed. B. Hauke, Gaganelis, Forman, Mark - Stahlbeton optimiert, Institut Bauen und Umwelt e.V., and DGNB e.V.). Ernst & Sohn, 39–44. ISBN: 9783433033357.

23 Papageorgiou, M., Leibold, M., and Buss, M. (2015). *Optimierung: Statische, dynamische, stochastische Verfahren für die Anwendung*. Berlin and Heidelberg: Springer Vieweg. ISBN: 9783662469361.

24 Belegundu, A.D. and Chandrupatla, T.R. (2014). *Optimization Concepts and Applications in Engineering*. New York: Cambridge University Press. ISBN: 0521878462.

25 Haftka, R.T. and Gürdal, Z. (1993). *Elements of Structural Optimization*. Dordrecht: Kluwer.

26 Arora, J.S. (2011). *Introduction to Optimum Design*. Amsterdam: Elsevier Academic Press.

27 Schumacher, A. (2013). *Optimierung mechanischer Strukturen: Grundlagen und industrielle Anwendungen*. Berlin and Heidelberg: Springer-Verlag. ISBN: 978-3-642-34699-6.

28 Harzheim, L. (2008). *Strukturoptimierung: Grundlagen und Anwendungen*. Frankfurt am Main.

2

Fundamentals of Reinforced Concrete (RC) Design

Key learnings after reading this chapter:

- What are the material properties of concrete and reinforcement?
- What are the basic design principles?
- How do RC structures bear loads?

2.1 Basic Principles

Concrete is the most popular building material in the world. This is mainly due to three reasons. First, its raw materials, namely cement, aggregates, and water, are widely available. Second, it offers very low production costs compared to other construction materials, for example, steel. Third, it provides the benefit of free shaping of structures and is thus highly versatile. While it exhibits excellent load-bearing properties under compression, its tensile strength is marginal at ca. 10% of its compressive strength. Therefore, the tensile strength is usually neglected in the design. In structural regions subject to tension, it must be strengthened by suitable reinforcement, which is statically activated as soon as the tensile strength is exceeded and concrete cracks appear (Figure 2.1a). These reinforcements may be of different shapes (bars, mats, and fibers) as well as of different materials (steel, alkali-resistant glass, and carbon) [1–4]. Figure 2.1b compares different types. However, steel bars are the most common. Consequently, reinforced concrete (RC) acts as composite material, where the concrete bears compressive stresses and the reinforcement is limited to structural regions undergoing tension. Forces between concrete matrix and embedded reinforcement are transmitted by a combination of frictional, adhesive, and mechanical bond. In the case of conventional steel rebars, for example, mechanical bond is achieved via the embossed ribs (Figure 2.1c) which ensure that concrete and reinforcement exhibit almost equal deformation since slippage is prevented.

The above description indicates that the load-bearing behavior of RC structures highly depends on whether they are cracked or not. These two conditions are typically referred to as state I (uncracked) and state II (cracked). The different load-bearing behavior in these two states will be illustrated subsequently by means of a RC beam enhanced by steel rebars in bending as illustrated in Figure 2.2 [5].

Optimization Aided Design: Reinforced Concrete, First Edition.
Georgios Gaganelis, Peter Mark, and Patrick Forman.
© 2022 Ernst & Sohn GmbH & Co.KG. Published 2022 by Ernst & Sohn GmbH & Co.KG.

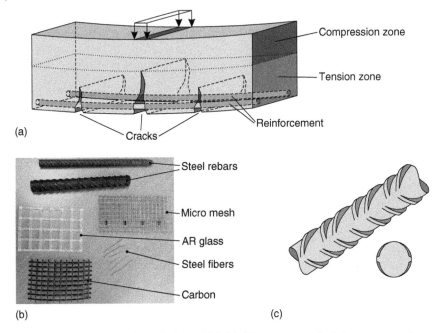

(a)

Compression zone

Tension zone

Reinforcement

Cracks

Steel rebars

Micro mesh

AR glass

Steel fibers

Carbon

(b)   (c)

Figure 2.1 (a) Load-bearing principle of RC, (b) different types of reinforcement, and (c) steel rebar with ribs.

As long as the tensile strength (f_{ct}) of the concrete is not exceeded at any point of the cross section, the beam remains uncracked $(\sigma_{ct} < f_{ct})$. It behaves like consisting of a homogeneous material. Both strains and stresses are linearly distributed over the cross-sectional height. In the absence of a normal force (pure bending), the neutral axis coincides with the center of gravity of the beam. The internal force flow is characterized by the principal stress trajectories.

As soon as the tensile strength of the concrete is exceeded, the first crack emerges and the beam slowly switches to state II. The transition between the states occurs gradually and ends only when all cracks have formed. In state II, the concrete merely bears the compressive stresses, whereas the rebars carry the tensile stresses at the cracks. The load transfer mechanisms are now based on the composite cross-section. Here, the cross section carries the bending moment via internal resulting compression (F_c) and tension forces (F_t). The former results from the stress distribution within the compression zone and the latter arise from the rebars. Although the strains are still distributed linearly over the cross section, the neutral axis does not necessarily coincides with the axis of gravity. In fact, the elongation state adjusts in such a way that the resulting internal forces $(F_c$ and $F_t)$ are in equilibrium with the external acting bending moment (M). The height of the compression zone decreases with increasing load and is approximately limited by the cracks. The compressive stresses of the concrete gradually become nonlinear. In the tensile zone, the stresses are concentrated on the reinforcement, since no loads can be transferred across the crack flanks. The stiffness decreases compared to state I, where it was equal to the elastic bending stiffness, what causes a disproportionate increase in deformations.

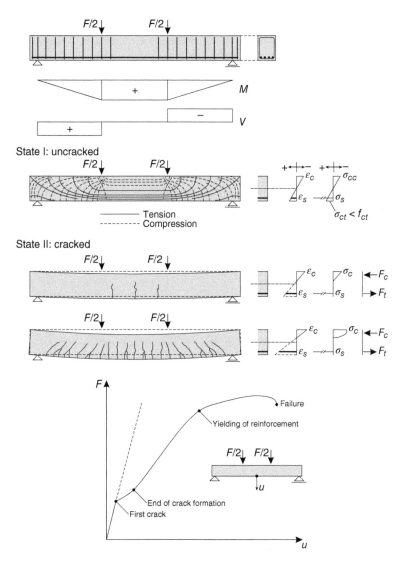

Figure 2.2 Load-bearing behavior of a bending beam in uncracked and cracked condition, according to [5].

If the bending moment grows further, the yield strength of the rebars is reached, which results in plasticizing. Beyond this point, only minor load increase is possible up to the point where the compression zone constricts to such an extent that the concrete crushes due to exceeding of the compressive strength.

2.2 Verification Concept

Three fundamental requirements are imposed for load-bearing structures: bearing capacity, serviceability, and durability [6, 7].

The **bearing** requirement implies that the structure must be able to bear all loads acting during its entire construction and life cycle. The various computational verifications must be carried out in the ultimate limit state (ULS). Due to safety reasons, failure announcement must be guaranteed, for example by excessive deformations and pronounced cracking. Moreover, ductility has to be ensured to avoid abrupt or brittle failure.

The **serviceability** limit state (SLS) verifications, on the other hand, guarantee the proper use of the structure. The checks include, for example, the limitation of excessive deformations and crack widths in SLS. However, exceeding the thresholds does not endanger the load-bearing capacity of the structure.

The **durability** requirement ensures the resistance of the structure to environmental influences and long-term effects that can damage the concrete matrix or the reinforcement (steel corrosion). Usually, no computational checks are conducted. Instead, durability is targeted by following rules for design and detailing and through complying with the ULS checks.

2.3 Safety Concept

For all verifications, it holds that the effect of action (E) must be lower than or equal to the resistance (R). Since material properties, geometrical dimensions, and acting loads are all subject to statistical fluctuations, these uncertainties have to be included in the limit state checks. The fluctuations are covered by stochastic means from which partial safety factors are derived for both the external loads and the internal resistances. In doing so, the former are computationally enlarged and the latter decreased:

$$\gamma_E \cdot E_k = E_d \leq R_d = R_k / \gamma_R \tag{2.1}$$

where E_k denotes the so-called characteristic effect of action (e.g. stress resultants), R_k is the characteristic internal resistance value, whereas $\gamma_E \geq 1.0$ and $\gamma_R \geq 1.0$ are the partial safety factors for the effect and resistance, respectively. The subscript k indicates so-called characteristic values whereas d denotes design quantities. Hence, the approach is twofold. On the one hand, there are characteristic values of actions and resistances. These typically constitute 5% quantiles of the assumed probability distribution, meaning that statistically they are only exceeded (actions) or undercut (resistances) in 5% of all cases. In structural engineering, a normal distribution is commonly used. On the other hand, there are the partial safety factors, which increase (actions) or reduce (resistances) the characteristic values to design quantities in order to ensure a sufficiently high safety level in the limit states.

Figure 2.3 shows the probability distribution of an action and its corresponding resistance as independent functions. The probability functions exhibit different means (μ_E, μ_R) and standard deviations (σ_E, σ_R). Due to the overlapping of the functions, a zone of calculative failure (gray) results, where the action is greater than the resistance. The failure probability can never become zero, since the normal distribution is defined between $-\infty$ and $+\infty$. Consequently, an absolute safety does

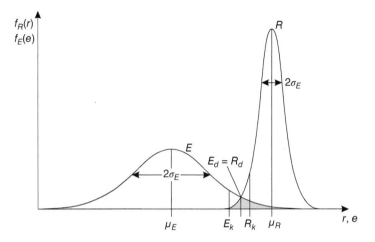

Figure 2.3 Probability distributions of an action (E) and a resistance (R), according to [5].

not exist. However, standards usually define an upper limit value of the tolerable operative probability of failure, which must be ensured in the limit states. From this, the safety factors to comply with the demanded safety level are computed. In Europe, for example, the safety requirements defined by the Eurocodes [6, 8] result in the partial safety factors $\gamma_C = 1.5$ for concrete, $\gamma_S = 1.15$ for reinforcing steel, and $\gamma_G = 1.35$ and $\gamma_Q = 1.5$ for permanent and variable loads, respectively.

2.4 Materials

2.4.1 Plain Concrete

Concrete is an artificial stone mainly consisting of cement, aggregates, water and, in certain cases, further additives. Thus, it is actually an inhomogeneous material, but its macroscopic mechanical behavior can be assumed homogeneous for the structural design.

The compressive strength of concrete (f_c) is determined in axial load tests as the maximum achieved value. Usually, it lies within the range of 20–50 MPa for normal strength concretes (NC) and goes beyond 50 and 120 MPa for high performance concrete (HPC) and ultra-high performance concrete (UHPC), respectively. A typical stress–strain curve for NC is given in Figure 2.4a. The curve is nonlinear, yet begins with a virtually linear elastic portion to about $0.4 f_c$. For loads exceeding $0.4 f_c$ microcrack formation sets in and the strains increase disproportionally. The micro cracks gradually merge into macro cracks, which causes the curve to flatten further until reaching the maximum compressive strength (f_c). Beyond this point, even the macro cracks merge which finally causes splitting of the test specimen. A softening behavior is observed until failure. The compressive strength is usually tested 28 days after concreting, since the development of strength is mostly finished by this time.

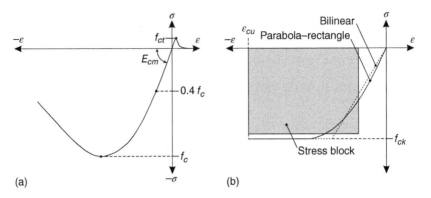

Figure 2.4 (a) Stress–strain diagram for concrete and (b) different material models for design.

Typical material models according to international standards [6, 7], which approximate the stress–strain diagram for the design, are depicted in Figure 2.4b. These range from nonlinear approaches, e.g. the parabola–rectangle diagram, to bilinear curves or simplified constant stress blocks. However, all consecutive models refer to the characteristic compressive strength (f_{ck}). The Young's modulus is determined in the linear branch as secant modulus (E_{cm}) and takes on values between about 30 000 and 40 000 MPa for NC. It is mainly influenced by the Young's modulus of the aggregates than by that of the cement. The Poisson's ratio is load-dependent, but a reasonable approximation is $v = 0.2$ in uncracked states.

Compared to its compressive strength, the axial tensile strength of concrete is very low and amounts to approximately 10% of the former. The tensile Young's modulus, however, is comparable to that under compression. In the design, this low tensile strength is usually neglected. Tensile stresses are, therefore, completely assigned to the embedded reinforcement.

The mechanical behavior of concrete described above refers to short-term effects. However, its behavior under long-term effects differs because time-dependent portions of deformation from creep and shrinkage are added. These portions can be multiple times higher than the short-term displacements. Particularly in the SLS, they must therefore be given due consideration. Furthermore, the compressive strength under long-term effects is also lower than under short-term loading. In current standards [6, 7], this effect is taken into account by computationally decreasing f_{ck} for the design.

2.4.2 Fiber-Reinforced Concrete (FRC)

During the mixing process, fibers can be added to the concrete in order to improve its post-cracking mechanical properties. This is due to the ability of the fibers to transmit tensile forces across the crack flanks (Figure 2.5a). Like conventional concrete, fiber-reinforced concrete (FRC) can also further be enhanced with additional bar or mesh-like reinforcement. In this case, the reinforcing effects are superimposed and the load-bearing capacity increases considerably.

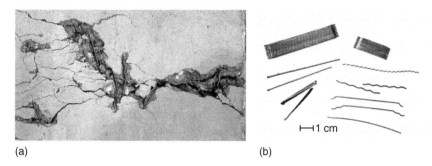

(a)   (b)

Figure 2.5 (a) Tensile force transmission across cracks and (b) various steel fiber types, according to [9].

Usually, fibers made of steel are used. Figure 2.5b gives an overview of some typical forms and shapes in practice. Besides steel, fibers for special applications are available which consist of synthetic materials (polypropylene), alkali-resistant glass, or carbon [10].

The post-cracking behavior of FRC scatters strongly and depends, among others, on the geometry, amount, and bonding behavior of the embedded fibers as well as on their orientation within the structure relative to the acting tensile stresses [10–14]. Under compression, FRC behaves similar to plain concrete. However, the fibers bridge the cracks and can thus provide for a more ductile post-cracking behavior. The higher the fiber content, the more the load can be carried (Figure 2.6a). The actual advantage of fiber concrete, however, becomes

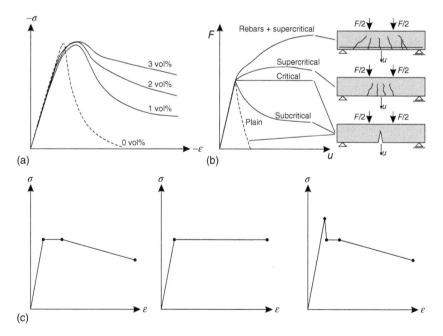

Figure 2.6 (a) Stress–strain relation of FRC under compression, (b) load-bearing behavior in bending, and (c) tensile stress–strain models for design, according to [12, 15, 16].

apparent under tensile stress, for example, in bending. Here, distinction is made between so-called subcritical and supercritical fiber contents, cf. Figure 2.6b. While in the former case the load decreases after cracking (strain softening), in the latter case the load can even be increased (strain hardening). In combination with bar reinforcement, the load-bearing capacity is increased further.

For the design, FRC can be treated like plain concrete via the stress–strain relationships given in Figure 2.4b under compression. For tension, different material models exist. Some of them, which originate from [16], are given in Figure 2.6c. They usually consist of stepwise linear branches.

2.4.3 Ultra-High Performance Concrete (UHPC)

UHPCs exhibit higher compressive strengths than covered by current standards, for example > C100/115 according to Eurocode 2 [6] or > C120 according to *fib* Model Code 2010 [7]. Their high strength arises from the very dense micro structure with extraordinary low pore content, which also provides excellent durability against environmental influences [17]. As a result, the cement matrix shows a smaller deficit in strength compared to the aggregates. Test specimens made of UHPC, therefore, exhibit straight cracks running through both cement matrix and aggregates, whereas in those made of NC the cracks only run through the former.

Under axial compression, UHPC shows an almost linear increase up to approximately $0.8f_c$. The curve then flattens and sudden, very brittle failure occurs after reaching f_c (Figure 2.7a). For this reason, UHPC can practically only be used if fibers are added to increase its ductility. In this case, like with FRC, however, the course of the curve shows significant fluctuation after reaching the compressive strength. Therefore, a simple linear constitutive model is conveniently used for the design

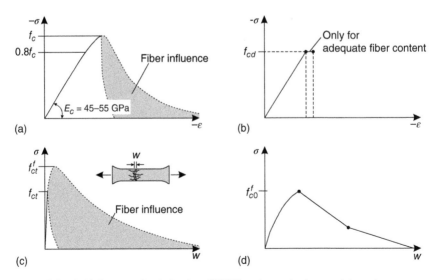

Figure 2.7 (a, b) Compressive behavior of UHPC and constitutive models and (c, d) stress–crack width relationship in tension and constitutive model, according to [17].

(Figure 2.7b). Alternatively, if a sufficiently ductile behavior is provably achievable by fiber addition, a bilinear approach with a short horizontal portion may also be applied [17]. The Young's modulus is higher in comparison to NC and is typically found within the range $E_c = 45\,000–55\,000$ MPa. The Poisson's ratio is comparable and can be assumed constant between $v = 0.18–0.20$.

Under axial tension UHPC behaves similarly to NC. If fibers are added, however, the post-cracking tensile strength (f_{ct}^f) and ductility are considerably improved. Figure 2.7c shows the stress with respect to the crack width (w) in tension. Also in this case, significant scattering of both the achievable post-cracking tensile strength as well as the curve shape is obvious. However, the fiber-induced tensile load-bearing capacity can be used in the design. Figure 2.7d presents a suitable three-part constitutive model according to [17].

2.4.4 Reinforcement

Reinforcement serves to compensate for the very low tensile strength of concrete after cracking. For economic reasons, it should be limited to structural areas that are subject to tensile stresses. Typically, ribbed reinforcing steels are used in the form of bars or mats. The material grades are commonly classified by their yield strength. Different types of reinforcing steel exist worldwide. In Germany, for example, only type B 500 according to DIN 488 [18] is used, which exhibits a characteristic yield strength of $f_{yk} = 500$ MPa. In contrast, in the United states, rebar yield strength of 420 MPa is common. However, generally it can be expected to be within the range 400–600 MPa.

Unlike concrete, reinforcing steel is characterized by an equal load-bearing behavior both in compression and tension (Figure 2.8a). The Young's modulus for all types lies between 200 000 and 210 000 MPa and the Poisson's ratio can be assumed to $v = 0.3$. After an initial linear elastic portion, plastification sets in ($>f_y$) and leads to a disproportionate increase in elongation. However, the load is further increased until failure (f_u), cf. Figure 2.8a. Bilinear approaches can be used as stress–strain material model for the design (Figure 2.8b), which may or may not incorporate the strain hardening behavior beyond yielding.

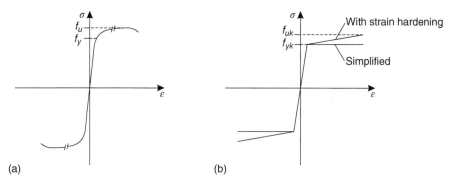

(a) (b)

Figure 2.8 (a) Stress–strain curve for reinforcing steel and (b) consecutive material models, according to [6, 7].

2.5 Load-Bearing Behavior

The load-bearing behavior of RC structures cannot be described comprehensively within this book since this would go beyond the scope. Instead, the brief descriptions following are limited to the topics addressed by the later introduced optimization aided methods for design. Essentially, the main focus lies on axial and biaxial bending with and without normal forces, and strut-and-tie models (STMs). Most importantly, standard-dependent topics, such as (semi-)empirical approaches for shear, torsion, or punching shear design, are explicitly not covered.

2.5.1 Bending Design

2.5.1.1 Fundamentals

In cracked concrete, the load-bearing behavior in bending becomes nonlinear, which leads to the superposition principle losing its validity. If besides the bending moment (M) also a normal force (N) is present, both actions must be taken into account simultaneously in the design, since a separate treatment leads to incorrect results. In the following, the expression "bending design" therefore also includes the normal force, if present.

The fundamental assumption is that the internal forces are linked to the axial cross-sectional stresses σ via equilibrium conditions. Following this, in the plane case then holds

$$N = \int_A \sigma dA \tag{2.2a}$$

$$M_y = \int_A \sigma z dA \tag{2.2b}$$

The internal forces thus result from the integration of the axial stresses (σ) over the cross-sectional area (A). The variable z denotes the vertical coordinate. In the spatial case, the equilibrium conditions must also account for the additional bending moment and extend to

$$N = \int_A \sigma dA \tag{2.3a}$$

$$M_y = \int_A \sigma z dA \tag{2.3b}$$

$$M_z = -\int_A \sigma y dA \tag{2.3c}$$

where y is the horizontal coordinate.

The theory of plane or spatial beams applies, namely the longitudinal extension x of the beam with respect to its two cross-sectional dimensions in y- and z-direction is significantly greater ($l \gg h$, $l \gg b$) [19]. Some further basic assumptions have to be made for a practical approach to bending design. First, Bernoulli's hypothesis applies, or in other words, no axial stress components arise from shear deformations. The strains ε form a plane that can be expressed as a linear function of the cross-sectional coordinates. In the two-dimensional case, it holds

$$\varepsilon = b_1 + b_2 z \tag{2.4}$$

where b_1 is the longitudinal (x-axis) elongation ε_x at the center of gravity and b_2 denotes the curvature κ_y with respect to the y-axis. The spatial case, on the other hand, requires the introduction of an additional summand referring to the y-direction:

$$\varepsilon = b_1 + b_2 z + b_3 y \tag{2.5}$$

where b_3 denotes the curvature κ_z related to the z-axis. Second, shear deformations are neglected. Third, rigid bond between concrete and embedded reinforcement is assumed, meaning that both exhibit the same strains in the same cross-sectional axes. Thus, slippage is excluded. Last, for all $j = 1,2, \dots, m$ materials that might compose a (multi-material) cross-section uniaxial stress–strain relations are given, cf. Section 2.4. They can be arbitrarily linear or nonlinear – as long as they stay physically reasonable – and include, for example, tension failure (compression only), plasticity, or ascending and descending branches:

$$
\begin{aligned}
\sigma_1 &= \sigma_1(\varepsilon) \\
\sigma_2 &= \sigma_2(\varepsilon) \\
&\vdots \\
\sigma_m &= \sigma_m(\varepsilon)
\end{aligned}
\tag{2.6}
$$

2.5.1.2 Equilibrium for Composite Sections

Equations (2.2) and (2.3) constitute the basis of the equilibrium for the plane and spatial case, respectively. The composite cross-section is composed of m materials. Each material M is assigned to an area of the cross-section, that is, A_1 consists of M_1, A_2 consist of M_2, and so on. The individual areas add up to the full cross-sectional area A:

$$A = \sum_{j=1}^{m} A_j \tag{2.7}$$

For solving the equilibrium equations, it is convenient to calculate the integrals for the individual areas separately. The stress integration then follows in general:

$$\int_A \sigma \, dA = \sum_{j=1}^{m} \int_{A_j} \sigma_j \, dA_j \tag{2.8}$$

Uniaxial Bending In the case of singly symmetrical cross-sections and plane loading from N and M_y, the curvature κ_z with respect to the second principal axis z is omitted. Thus, a plane strain state is given. The strains lead to stresses at every point of the cross-section, the magnitude of which depends on the stress–strain relationship of the material. Figure 2.9 shows this principle for a plane composite cross-section of $m = 3$ materials M_1 to M_3. Material changes lead to abrupt changes

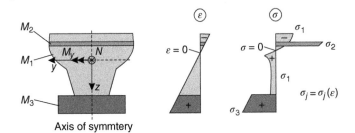

Axis of symmtery

Figure 2.9 Symmetric composite section under plane loadings, layered section of $m = 3$ materials M_1, M_2, and M_3, distribution of strains ε, distribution of stresses σ.

in the stress curves, whereas, in contrast, the strains are linear and continuous. In order to formulate the equilibrium, it is reasonable to differentiate between materials, meaning to first integrate over the single areas A_j and then form the sum of all individual proportions. In doing so, the equilibrium equations take the form:

$$N = \int_{A_1} \sigma_1 dA_1 + \int_{A_2} \sigma_2 dA_2 + \cdots + \int_{A_m} \sigma_m dA_m \tag{2.9a}$$

$$M_y = \int_{A_1} \sigma_1 z dA_1 + \int_{A_2} \sigma_2 z dA_2 + \cdots + \int_{A_m} \sigma_m z dA_m \tag{2.9b}$$

This corresponds to Eq. (2.2) in a material-differentiating notation.

Biaxial Bending In the spatial case, the described principle remains, but an additional equilibrium equation for the bending moment M_z is added. Equation (2.9) is thus extended by a further integral equation in accordance with Eq. (2.3):

$$N = \int_{A_1} \sigma_1 dA_1 + \int_{A_2} \sigma_2 dA_2 + \cdots + \int_{A_m} \sigma_m dA_m \tag{2.10a}$$

$$M_y = \int_{A_1} \sigma_1 z dA_1 + \int_{A_2} \sigma_2 z dA_2 + \cdots + \int_{A_m} \sigma_m z dA_m \tag{2.10b}$$

$$M_z = \int_{A_1} \sigma_1 y dA_1 + \int_{A_2} \sigma_2 y dA_2 + \cdots + \int_{A_m} \sigma_m y dA_m \tag{2.10c}$$

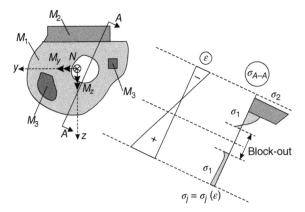

Figure 2.10 Arbitrary composite section under spatial loadings, section of $m = 3$ materials M_j ($j \in [1, m]$), distribution of strains ε and stresses σ in section A-A.

Again, only a differentiation according to the individual materials is made, namely an additive separation referring to the portions of all individual areas. The strain and stress distributions change over both the y- and z-direction, thus a representations in line plots are valid only for discrete sections. Figure 2.10 shows this for a section A-A, which also contains a block-out where no stresses can be taken.

RC or FRC sections For RC or FRC, two materials must be distinguished. This is first (M_1) concrete or FRC and second (M_2) the steel rebars. In order to remain compliant with common conventions, the materials are not numbered, but denoted with the indices c (concrete or FRC) and s (reinforcing steel). Concrete and FRC can be treated in the same way, since in the case of the latter merely the capability of tensile stress transmission must be additionally accounted for, cf. Section 2.4.2. The load-bearing behavior in compression, however, is equal, whereas the nonlinear constitutive law remains.

The steel area of a reinforcing bar is small compared to the cross-sectional dimensions and can therefore be idealized as point-like. The strains, stresses, and cross-sectional coordinates are determined at its center. Moreover, integration over the small area is not necessary. Consequently, two proportions arise, on the one hand, integrals over the concrete area A_c and, on the other hand, sums of the n individual reinforcement points $A_{s1}, A_{s2}, ..., A_{sn}$. The concrete area A_c would be very difficult to integrate as net area by subtracting the reinforcement points. It is therefore convenient to first integrate the concrete stresses over the full cross-sectional area A and then subtract the incorrectly applied concrete stresses in the sums of the steel portions. Following this approach, the equilibrium equations yield

$$N = \int_{A_c} \sigma_c dA_c + \sum_{i=1}^{n} \sigma_{si} A_{si} = \int_A \sigma_c dA + \sum_{i=1}^{n} \left(\sigma_{si} - \sigma_{ci} \right) A_{si} \tag{2.11a}$$

$$M_y = \int_{A_c} \sigma_c z dA_c + \sum_{i=1}^{n} \sigma_{si} A_{si} z_i = \int_A \sigma_c z dA + \sum_{i=1}^{n} \left(\sigma_{si} - \sigma_{ci} \right) A_{si} z_i \tag{2.11b}$$

$$M_z = \int_{A_c} \sigma_c y dA_c + \sum_{i=1}^{n} \sigma_{si} A_{si} y_i = \int_A \sigma_c y dA + \sum_{i=1}^{n} \left(\sigma_{si} - \sigma_{ci} \right) A_{si} y_i \tag{2.11c}$$

The subtraction can, of course, also be omitted if gross cross-sectional values are used for simplicity. This is justified in the case of low geometric reinforcement ratios.

Figure 2.11 shows an arbitrary RC cross-section under spatial loading with the corresponding strain and stress distributions at a section. As is well known, the steel stresses significantly exceed the concrete stresses at equal strain values. Besides its bearing capacity in compression, the concrete in Figure 2.11 is also assigned a tensile stress capacity, resulting, for example, from the smeared effect of fibers. This serves to illustrate the generality of material laws, which can represent "compression only" behavior in the same way as tensile and compressive behaviors.

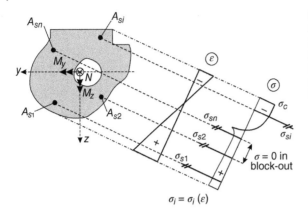

Figure 2.11 Arbitrary RC section with *n* reinforcement points A_{si}, strain (ε), and stress (σ) distributions including tensile-bearing capacities for concrete (e.g. FRC).

2.5.2 Strut-and-Tie Models (STMs)

In the case of structures or partial sections where the assumption of Bernoulli's theory is no longer valid, they are referred to as discontinuity regions (D-regions). Here, the use of conventional design approaches assuming plane strain distribution is prohibited.

A D-region may be due to geometric (Figure 2.12a) or static discontinuities (Figure 2.12b). In the uncracked state, linear elastic approaches are still applicable, for instance, Hooke's law. In contrast, the cracked state requires the application of so-called strut-and-tie models (STMs) for design. It should be emphasized that STMs are actually not strictly limited to D-regions but may also be employed to represent the load-bearing mechanisms of B-regions (Bernoulli regions) likewise. Figure 2.12c illustrates such a unified design approach proposed, for instance, in [20].

STMs consist of compression struts and tension ties, which are connected in nodes. The struts and ties merge the actual principle stress distribution to straight resulting forces. The curvature of the stresses is bundled into the nodes. The basic concept for RC design also applies in this case: the concrete is assigned to the struts and the reinforcement is assigned to the ties. The design procedure with STMs consists of several steps. First, the B- and D-regions of a structure must be identified. The former are covered with well-known standard models like the bending theory introduced in Section 2.5.1. For the latter, the internal load transfer and a corresponding STMs must be found. Finite element (FE) stress calculations may provide orientation (Figure 2.13). Subsequently, the individual struts and ties must be dimensioned. The struts represent the force transmission in concrete. Consequently, verification is based on the effective strength (f_{cd}^*) of the concrete strut:

$$\frac{C_{d,i}}{A_{c,i}} \leq f_{cd}^* \tag{2.12}$$

where $C_{d,i}$ is the force of the strut i and $A_{c,i}$ is its cross-sectional area. The effective compressive strength depends on the presence of cracks and reinforcement that cross the strut and decrease its load-bearing capacity. The ties carry tensile forces and must coincide, for instance, with the rebar layer axes in case of conventional

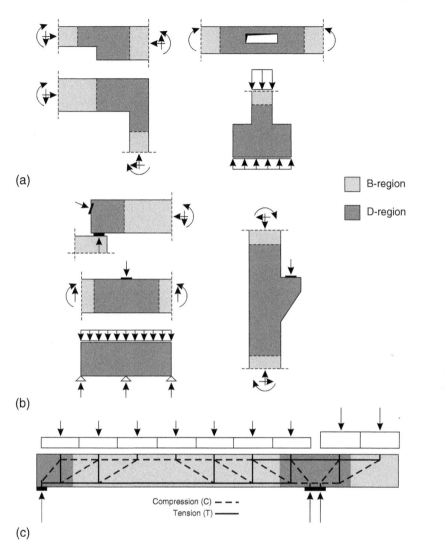

Figure 2.12 (a) Geometrical discontinuities, (b) statistical and/or geometrical discontinuities, and (c) uniform design approach using STMs for B- and D-regions, according to [20].

bar reinforcement. In doing so, each tie i must meet the verification

$$T_{d,i} \leq A_{s,i} f_{yd} \tag{2.13}$$

In Eq. (2.13), $T_{d,i}$ is the tie force, $A_{s,i}$ is the cross-sectional area of the reinforcement representing the tie and f_{yd} denotes the design yield strength.

STMs must meet certain requirements in order to be qualified for design purposes. The most important is the equilibrium of forces. In this context, the lower bound theorem of plasticity serves as the theoretical basis. However, since concrete can only perform limited plastic deformations, the STMs must be specified in such a way

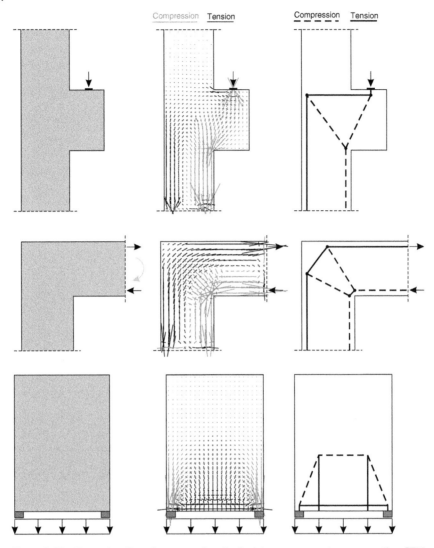

Figure 2.13 Exemplary D-regions, associated principle stresses and corresponding STMs.

that its deformation limit is not exceeded before the assumed stress state is reached. This aspect becomes particularly important for highly utilized structures in ULS. However, this requirement is met if the STMs is oriented toward the elastic stress distribution in the uncracked state. In this way, excessive cracking is prevented. In many cases, multiple STMs are applicable to approximate the stress distribution in state I. The question thus arises based on which criterion STMs can be compared to each other.

Generally, loads preferably follow a direct path to the supports with minimum deformations. STMs with fewer struts and ties should therefore be preferred over others, which leads to the following criterion:

$$\sum_i N_i L_i \varepsilon_{m,i} \rightarrow \min \tag{2.14}$$

where N_i is the strut or tie force, L_i is its respective length, and $\varepsilon_{m,i}$ is the mean strain of member i. However, the reinforcement material can usually exhibit higher strains than concrete. Thus, a more sophisticated criterion is to prefer STMs with the fewest and shortest ties [20].

Some problems may arise when attempting to use STMs for the design of RC structures. As mentioned above, developing suitable models requires detailed knowledge of the flow of forces. For this purpose, a stress analysis based on FE computations can be conducted. An alternative approach is the so-called "load path method" as described in [20]. In some particularly complicated cases, combining both methods may even be required. However, a certain amount of experience is demanded from the designer.

References

1 Curbach, M. and Jesse, F. (2009). Specifications and application of textile reinforced concrete (TRC). *Beton- und Stahlbetonbau* 104 (1): 9–16.

2 Brückner, A., Ortlepp, R., and Curbach, M. (2006). Textile reinforced concrete for strengthening in bending and shear. *Materials and Structures* 39 (8): 741–748. https://doi.org/10.1617/s11527-005-9027-2.

3 Schladitz, F., Frenzel, M., Ehlig, D., and Curbach, M. (2012). Bending load capacity of reinforced concrete slabs strengthened with textile reinforced concrete. *Engineering Structures* 40: 317–326. https://doi.org/10.1016/j.engstruct.2012.02.029.

4 Schumann, A., May, M., and Curbach, M. (2018). Rebars made of carbon fibres for the use in civil engineering: Part 1: Fundamental material characteristics. *Beton- und Stahlbetonbau* 113 (12): 868–876.

5 Zilch, K. and Zehetmaier, G. (2010). *Bemessung im konstruktiven Betonbau.* Berlin and Heidelberg: Springer.

6 EN 1992-1-1 (2011). *Eurocode 2: Design of concrete structures - Part 1-1: General rules and rules for buildings.*

7 Féderation Internationale du Béton (2013). *fib Model Code for Concrete Structures 2010.* Ernst & Sohn. ISBN: 9783433604090.

8 EN 1990 (2010). *Eurocode 0: Basis of structural design.* Brussels: Comité Européen de Normalisation (CEN).

9 Look, K., Oettel, V., Heek, P. et al. (2020). Bemessen mit Stahlfaserbeton. In: *Beton-Kalender 2021* (ed. K. Bergmeister, F. Fingerloos, and J.-D. Wörner), 797–874. Berlin: Ernst & Sohn. ISBN: 978-3-433-03301-2.

10 Holschemacher, K., Dehn, F., and Klug, Y. (2011). Grundlagen des Faserbetons. In: *Beton-Kalender 2011* (ed. K. Bergmeister, F. Fingerloos and J.-D. Wörner), 19–88. Ernst & Sohn. ISBN: 9783433601013.

11 Gödde, L. (2013). Numerische Simulation und Bemessung von Flächentragwerken aus Stahlfaserbeton sowie stahlfaserverstärktem Stahl- und Spannbeton. PhD thesis. Bochum: Ruhr University Bochum.

12 Heek, P. (2018). Modellbildung und numerische Analysen zur Ermüdung von Stahlfaserbeton. PhD thesis. Bochum: Ruhr University Bochum.

13 Falkner, H. and Grunert, J.-P. (2011). Faserbeton. In: *Beton-Kalender 2011* (ed. K. Bergmeister, F. Fingerloos, and J.-D. Wörner), 1–17. Ernst & Sohn. ISBN: 9783433601013.

14 Strack, M. (2007). Modellbildung zum rissbreitenabhängigen Tragverhalten von Stahlfaserbeton unter Biegebeanspruchung. PhD thesis. Bochum: Ruhr University Bochum.

15 Fanella, D. and Naaman, A. (1985). Stress-strain properties of fiber reinforced mortar in compression. *ACI Journal Proceedings* 82 (4). https://doi.org/10.14359/10359.

16 Deutscher Ausschuss für Stahlbeton (2012). *DAfStb-Richtlinie Stahlfaserbeton.* Beuth, Berlin.

17 Fehling, E., Schmidt, M., Walraven, J.C. et al. (2014). *Ultra-High Performance Concrete UHPC: Fundamentals, Design, Examples.* Berlin: Ernst & Sohn. ISBN: 9783433604069.

18 DIN 488-1 (2009). *Reinforcing steels - Part 1: Grades, properties, marking.*

19 Krätzig, W.B., Başar, Y., and Wittek, U. (1997). *Tragwerke.* Berlin: Springer-Lehrbuch, Springer. ISBN: 3-540-62440-6.

20 Schlaich, J., Schäfer, K., and Jennewein, M. (1987). Toward a consistent design of structural concrete. *PCI Journal* 32 (3): 74–150.

3

Fundamentals of Structural Optimization

Key learnings after reading this chapter:

- How is a structural optimization problem defined and what are its components?
- How can a structural optimization problem be practically studied?
- What approaches exist to solve a typical structural optimization problem?

3.1 Structural Optimization Approaches

3.1.1 General Procedure

The focus of this textbook lies on the numerical optimization methods for reinforced concrete (RC) structures. Therefore, the essential basics of such structural optimization approaches based on numerical techniques will be introduced below.

First, the structure to be studied has to be implemented into a simulation model that can reproduce its mechanical behavior with sufficient accuracy. The *simulation model* may consist of a set of analytical equations or, alternatively, it may include a numerical approximation approach, for instance, the finite element method (FEM) [1–3]. Input variables define its geometry, the material behavior, and all boundary conditions. Furthermore, a set of *design variables* is introduced, which link the simulation model with the optimization problem. The task now is to determine the design variable values in such a way that the *objective function* of the optimization problem is minimized, while at the same time eventual *constraints* are respected.

However, structural optimization problems are usually complex, hence the design variables have to be updated in several iterations. In doing so, the system responses (output variables) are determined for each iteration step. The design variables are updated on this basis. This, in turn, defines a new set of input variables for the subsequent iteration. The new system responses are calculated and the design variables are adjusted again. The optimization procedure is repeated until either:

Optimization Aided Design: Reinforced Concrete, First Edition.
Georgios Gaganelis, Peter Mark, and Patrick Forman.
© 2022 Ernst & Sohn GmbH & Co.KG. Published 2022 by Ernst & Sohn GmbH & Co.KG.

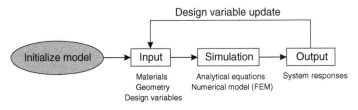

Figure 3.1 Basic procedure of a structural optimization.

- the global minimum is found, or
- the iterations are stopped manually, or
- a predefined convergence criterion is met.

Figure 3.1 illustrates the basic procedure.

3.1.2 Classification of Methods

The various structural optimization approaches can be roughly divided into three categories, which, however, may also overlap. Yet it is reasonable to distinguish between *topology optimization*, *shape optimization*, and *sizing* [4–10].

Topology optimization [9, 11] is the most versatile method, because inner and outer design of a structure are optimized at the same time. This is done by iteratively redistributing a limited amount of material within a given design space, which sequentially changes the structure's topology. Here, the design variables determine

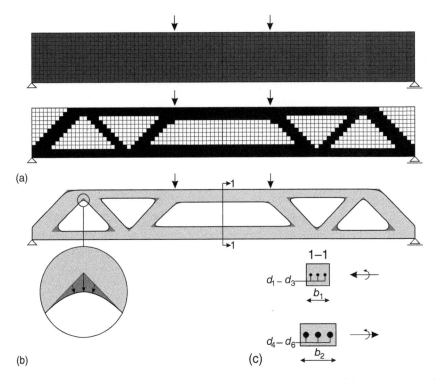

Figure 3.2 (a) Topology optimization, (b) shape optimization, and (c) sizing on the example of a beam in four-point bending.

the stiffness of structural regions and thereby control whether these regions are assigned material or not. This leads to a structure whose design follows the flux of forces. Figure 3.2a illustrates a topological optimization of a beam in four-point bending, which is converted into a truss structure through material distribution.

Shape optimization methods [6, 7, 10, 12, 13] aim at optimizing the outer edge of a structure but do not introduce any new void areas therein. The objective could be, for example, to compensate for stress concentrations or adjusting the structural height to the bending moment profile in order to minimize the deflection. Taking the topological optimized truss structure as an example, shape optimization is used in Figure 3.2b to reduce notch stresses at the borders.

Chapter 4: Identification of structures

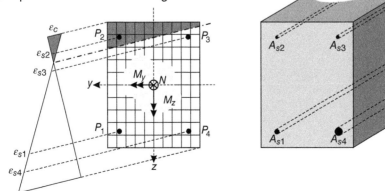

Chapter 5: Internal force flow

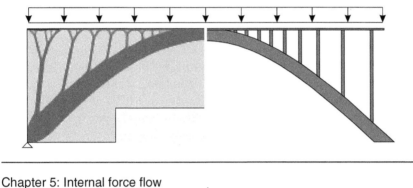

Chapter 6: Cross-section design

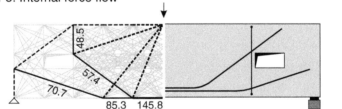

Figure 3.3 Organization of the approaches within this book.

Finally, discrete dimensions of the individual structural components can be optimized by sizing [14–16]. For example, cross-sectional areas and rebar diameters can be adjusted in order to increase utilization and decrease reinforcement amounts (Figure 3.2c). In the given example, bar reinforcements specified by their respective diameter d_1 to d_6 are added to the compressive and tensile flanges at the top and bottom to enhance the concrete and cover the tensile forces, respectively. The existence of the former is usually undesirable since it is uneconomical, but it serves as a general representation of an according sizing problem.

Deviating from the upper methodical-oriented categorization, the classification in this book follows a more practice-oriented motivation. Figure 3.3 gives an overview, starting with the identification of the global structure design, continuing with optimizing the internal load transmission and finally ending with detailed cross-sectional checks.

The organization thus ranges from a macro to a micro scale of the structural design and employs adequate optimization approaches for each of the purposes. For the (preliminary) design stage, topology optimization approaches are presented to identify an optimized overall shape of a structure (Chapter 4). Next, the internal force flow is revealed also by means of topology optimization. This time, however, based on a given overall structure design, optimized strut-and-tie models (STMs) are sought for determining minimum reinforcement layouts (Chapter 5). Finally, sizing is used to solve the challenging issue of determining the required reinforcement quantities of structural components exposed to biaxial bending (Chapter 6). Moreover, cross-sections are redesigned in the geometry and material usage. The aim is to ensure equilibrium of forces while minimizing the rebar cross-sectional areas. The geometry can be arbitrary, even asymmetrical, and enhanced material behavior, e.g. by means of steel fibers, can also be taken into account.

3.2 Problem Statement

An optimization problem [8, 10, 17–21] takes the general form:

$$
\begin{aligned}
\text{find: } & \mathbf{x} = [x_1, x_2, \ldots, x_N]^\mathsf{T} \\
\text{such that: } & f(\mathbf{x}) \to \min_{\mathbf{x}} \\
\text{subject to: } & g_j(\mathbf{x}) \leq 0, \qquad j \in [1, m] \\
& h_l(\mathbf{x}) = 0, \qquad l \in [1, q] \\
& \mathbf{x}^L \leq \mathbf{x} \leq \mathbf{x}^U
\end{aligned}
\tag{3.1}
$$

Here, $\mathbf{x}$ is a vector containing all N design variables, whereas $\mathbf{x}^L$ and $\mathbf{x}^U$ contain all their lower and upper bounds, respectively, $f(\mathbf{x})$ is the objective function, $g_j(\mathbf{x})$ is a set of r inequality constraints, and $h_l(\mathbf{x})$ is a set of q equality constraints. The constraints are given in the standard form, i.e. lower or equal zero. The objective function is typically formulated as minimization task. If instead the maximum is sought, it can easily be transformed [6] to

$$
\max(f(\mathbf{x})) = \min(-f(\mathbf{x}))
\tag{3.2}
$$

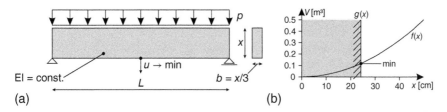

Figure 3.4 (a) Simple optimization problem and (b) objective function, inequality constraint, and optimal solution.

In any case, a solution found to the optimization problem is only permissible if the constraints are not violated.

For an introduction, consider the single span beam with rectangular cross section, which is loaded by $q = 5$ kN/m in Figure 3.4a. The span width amounts to $L = 6$ m. The structural height x is sought for a maximum deflection at midspan of 3 cm, while simultaneously minimizing the amount of material. The cross-sectional area is constant over the beam length and its width is given to $b = x/3$. Hence, the height is the only design variable, thus it is denoted by x. The objective function can be formulated as a minimization task of the volume:

$$f(x) = bxL = \frac{x^2 L}{3} \rightarrow \min_{x} \qquad (3.3)$$

For homogeneous material, the deflection at midspan reads

$$u(x) = \frac{5}{384EI} qL^4 \qquad (3.4)$$

where I denotes the moment of inertia. With $I = bx^3/12$ and a Young's modulus of $E = 30\ 000$ MPa, the minimum height which ensures the maximum allowable deflection of 3 cm is computed to

$$\sqrt[4]{\frac{5 \times 36 \times qL^4}{384 \times E \times 0.03}} = 0.241 \text{ m} \leq x \qquad (3.5)$$

The inequality constraint in standard form consequently reads

$$g(x) = 0.241 \text{ m} - x \leq 0 \qquad (3.6)$$

An equality constraint does not exist in this example. Figure 3.4b reveals that the optimal height equals $x^* = 24.1$ cm, i.e. it is located on the edge of the permissible solution space, which is limited by the constraint function. Thus, the minimum material amount, which results for maintaining the maximum beam deformation, yields $V = 0.12$ m^3.

Typically, however, structural optimization problems are not so easy to solve. The optimum must therefore usually be approached in several iterations by numerically updating the design variables step by step. For this purpose, an increment Δx is added to the design variable vector x at the end of each iteration $k - 1$:

$$x^{(k)} = x^{(k-1)} + \Delta x^{(k-1)} \qquad (3.7)$$

The approaches for determining $\Delta\mathbf{x}$ are numerous but can roughly divided into three categories [8, 10, 17–21].

In case of the *gradient descent methods*, the search direction and thus the increment are specified by the negative gradient of the objective function with respect to the constraints. An appropriate illustrative representation is that of a blind mountaineer who wants to reach the valley and therefore uses the steepest descent as orientation. The constraints represent insurmountable fences. Figure 3.5 illustrates that, however, the found solution strongly depends on the starting point. While this is irrelevant for convex problems that have only one minimum (=global optimum), there is high risk of reaching a local optimum for non-convex problems, where several minima exist. Methods to circumvent this issue even with non-convex problems are, for example, to calculate several solutions for different starting points and then select the one that leads to the lowest objective function value. However, even in this case, there is no guarantee that the global minimum is actually found.

The second group of approaches is known as *global optimization methods*. They rely on a similar strategy as the gradient descent methods but are enhanced by stochastic elements in order to counteract the issue of converging toward local minima. They thus correspond to the descent of a blind mountaineer, who generally follows the steepest path, but occasionally changes direction randomly.

The third group includes approaches based on *optimality criteria/conditions* (OC). OC refers to the mathematical description of a system state that is considered optimal from experience or observation. From this, a heuristic update scheme is developed which evolves the design variables toward the desired system state. OC can be either empirical or alternatively derived from the Lagrange function (Section 3.3) of the optimization problem.

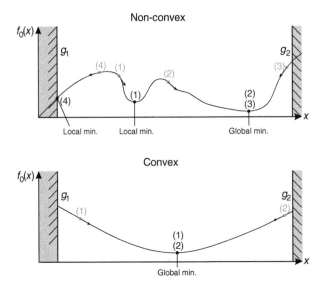

Figure 3.5 Starting point dependence of the solutions using gradient descent methods.

3.3 Lagrange Function

For a more efficient mathematical evaluation of the optimization problem, it is useful to combine the objective and the constraint functions into one single expression. The result of this merging is called *Lagrange function*. In the optimum, which is typically indicated by a superscript asterisk ($*$), the Lagrange function of the general optimization problem stated in Eq. (3.1) reads

$$L(\mathbf{x}^*, \lambda^*, \gamma^*) = f(\mathbf{x}^*) + \sum_{j=1}^{m} \lambda_j^* g_j(\mathbf{x}^*) + \sum_{l=1}^{q} \gamma_l^* h_l(\mathbf{x}^*) \tag{3.8}$$

Here, $\lambda_j \leq 0$ and $\gamma_l \leq 0$ are the *Lagrange multipliers*. Generally, they are zero and only take nonzero values when the respective constraint is active [7]. The Lagrange function exhibits the property of becoming stationary in the optimum, which means that its partial derivatives with respect to both the design variables and the Lagrange multipliers become zero [6, 7]. The derivatives are also known as *Karush–Kuhn–Tucker* (KKT) conditions [22]. They are given as follows:

$$\frac{\partial L(\mathbf{x}^*, \lambda^*, \gamma^*)}{\partial x_e} = \frac{\partial f(\mathbf{x}^*)}{\partial x_e} + \sum_{j=1}^{m} \lambda_j^* \frac{\partial g_j(\mathbf{x}^*)}{\partial x_e} + \sum_{l=1}^{q} \gamma_l^* \frac{\partial h_l(\mathbf{x}^*)}{\partial x_e} = 0$$

$$\frac{\partial L(\mathbf{x}^*, \lambda^*, \gamma^*)}{\partial \lambda_j} = g_j(\mathbf{x}^*) = 0 \tag{3.9}$$

$$\frac{\partial L(\mathbf{x}^*, \lambda^*, \gamma^*)}{\partial \gamma_l} = h_l(\mathbf{x}^*) = 0$$

$$\lambda_j g_j(\mathbf{x}^*) = 0$$

for all $e \in [1, N]$. The second and third expressions indicate that the constraints are met in the optimum. The fourth is known as *switching condition*. The KKT conditions are only necessary conditions. In order to clearly determine whether the optimum is a minimum, maximum, or saddle point, they must additionally be supplemented by sufficient conditions, in other words it has to be examined if the corresponding Hesse matrix is positive definite, negative definite, or indefinite.

For constrained optimization problems, such as those usually encountered in structural optimization, the stationarity of the Lagrange function indicates a saddle point [23–25]. As shown in Figure 3.6, a saddle point exists, if in the stationary

Figure 3.6 Saddle point of a Lagrange function containing one inequality constraint.

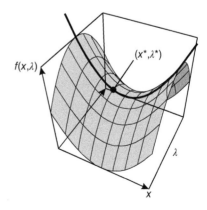

point, the Lagrange function becomes minimal with respect to the design variables and maximal with respect to the Lagrange multipliers:

$$L(\mathbf{x}^*, \lambda, \gamma) \leq L(\mathbf{x}^*, \lambda^*, \gamma^*) \leq L(\mathbf{x}, \lambda^*, \gamma^*) \tag{3.10}$$

Furthermore, the constraints must be satisfied. Thus, $L(\mathbf{x}^*, \lambda^*, \gamma^*) = f(\mathbf{x}^*)$ follows, i.e. the Lagrange function equals the objective function in the optimum.

3.4 Sensitivity Analysis

The partial derivatives of the objective and constraint functions with respect to the design variables are called *sensitivities* of the optimization problem:

$$
\begin{aligned}
\nabla f(\mathbf{x}) &= \left[\frac{\partial f(\mathbf{x})}{\partial x_1}, \dots, \frac{\partial f(\mathbf{x})}{\partial x_N} \right]^{\mathsf{T}}, \\
\nabla g_j(\mathbf{x}) &= \left[\frac{\partial g_j(\mathbf{x})}{\partial x_1}, \dots, \frac{\partial g_j(\mathbf{x})}{\partial x_N} \right]^{\mathsf{T}}, \quad j \in [1, m] \\
\nabla h_l(\mathbf{x}) &= \left[\frac{\partial h_l(\mathbf{x})}{\partial x_1}, \dots, \frac{\partial h_l(\mathbf{x})}{\partial x_N} \right]^{\mathsf{T}}, \quad l \in [1, q]
\end{aligned}
\tag{3.11}
$$

In other words, sensitivities are function gradients and contain information about the influence the design variable changes. Computing them is usually decisive for the total duration of the optimization, because it is one of the most expensive tasks [10, 12, 26–28]. Generally, they can be determined in two ways: numerically and (semi-)analytically. Both are briefly discussed hereafter.

3.4.1 Numerical Approach

The sensitivities can be approached very easily using the finite difference method. In the case of forward differentiation, the derivative of a function depending on N design variables with respect to x_e reads

$$\frac{\partial f(\mathbf{x})}{\partial x_e} \approx \frac{\Delta f(\mathbf{x})}{\Delta x_e} = \frac{f(x_1, \dots, x_e + \Delta x_e, \dots, x_N) - f(x_1, \dots, x_e, \dots, x_N)}{\Delta x_e} \tag{3.12}$$

In other words, the function value difference for a small change of a design variable is evaluated and divided by this change. For a function that depends only on a single variable, it can be represented as the slope of the triangle formed between these two points as depicted in Figure 3.7. Consequently, with N design variables,

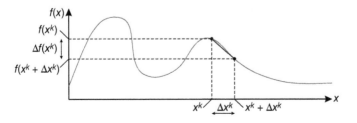

Figure 3.7 Forward differentiation for a function dependent on one variable.

N additional functional analyses must be performed for each optimization loop in order to determine all sensitivity values. Hence, the numerical approach is very expensive. It should only be used if the sensitivities cannot be determined in a more efficient way, that is by analytical methods.

3.4.2 Analytical Approach

Due to the direct reference to the later presented optimization techniques, the (semi-)analytical approach will be described using the FEM, which is often applied as the simulation model in structural optimization. In this case, starting point is its basic equation:

$$\mathbf{K}(\mathbf{x})\mathbf{U} = \mathbf{F} \tag{3.13}$$

where $\mathbf{K}$ is the global stiffness matrix, which depends on the vector $\mathbf{x}$ containing all design variables, and $\mathbf{U}$ and $\mathbf{F}$ are the global displacement and load vector, respectively. Assume that the optimization problem's objective function is a system response, for example, the mean structural compliance (c, inverse of stiffness), which is typically used in topology optimization approaches. For c then holds

$$c\left(\mathbf{x}, \mathbf{U}\left(\mathbf{x}\right)\right) = \mathbf{F}^{\mathsf{T}}\mathbf{U}(\mathbf{x}) = \mathbf{U}(\mathbf{x})^{\mathsf{T}}\mathbf{K}(\mathbf{x})\mathbf{U}(\mathbf{x}) \tag{3.14}$$

Obviously, the compliance depends on the displacements, which in turn depend on the design variables. In order to calculate the total differential, the chain rule must be employed and leads to

$$\frac{dc\left(\mathbf{x}, \mathbf{U}\left(\mathbf{x}\right)\right)}{dx_e} = \frac{\partial c\left(\mathbf{x}, \mathbf{U}\left(\mathbf{x}\right)\right)}{\partial x_e} + \sum_{j=1}^{N_u} \frac{\partial c\left(\mathbf{x}, \mathbf{U}\left(\mathbf{x}\right)\right)}{\partial u_j} \frac{\partial u_j}{\partial x_e} \tag{3.15}$$

with N_u being the number of degrees of freedom and u_j the respective displacement value. The partial derivatives of c with respect to the design variables can be obtained directly, since generally the relationship between the stiffness matrix and the design variables is known:

$$\frac{\partial c\left(\mathbf{x}, \mathbf{U}\left(\mathbf{x}\right)\right)}{\partial x_e} = \mathbf{U}(\mathbf{x})^{\mathsf{T}}\frac{\partial \mathbf{K}(\mathbf{x})}{\partial x_e}\mathbf{U}(\mathbf{x}) \tag{3.16}$$

In contrast, there is no explicit relation between the displacements and the design variables, so calculating the terms $\partial u_j/\partial x_e$ is particularly difficult. Two approaches are known to provide solution to this problem: the direct and the adjoint method [6, 10, 29]. The adjoint method will be discussed in more detail here, since it is very advantageous for the topology optimization approaches presented in the book involving many design variables and comparatively few constraints.

To promote readability, the dependencies on $\mathbf{x}$ are omitted in the notation hereafter. First, a zero term is determined from the basic FEM equation:

$$\mathbf{K}\mathbf{U} - \mathbf{F} = 0 \tag{3.17}$$

Even after multiplying the equation by an arbitrary factor, the zero character of each individual row remains unchanged:

$$\psi^{\mathsf{T}}\left(\mathbf{K}\mathbf{U} - \mathbf{F}\right) = 0 \tag{3.18}$$

Here, ψ is a vector containing M arbitrary multipliers ψ_j. The sum of all M zero-terms is then added to Eq. (3.14), thus altering the expression but not the value of the compliance:

$$c = c + \sum_{j=1}^{M} \psi_j \left(\sum_{k=1}^{M} K_{jk} u_k - F_j \right) \tag{3.19}$$

K_{jk}, u_k, and F_j are the respective entries of the $M \times M$ symmetrical global stiffness matrix, displacement, and load vectors, respectively. The total differential then reads

$$\frac{dc\,(\mathbf{x}, \mathbf{U}(\mathbf{x}))}{dx_e}$$

$$= \frac{\partial c}{\partial x_e} + \sum_{k=1}^{M} \frac{\partial c}{\partial u_k} \frac{\partial u_k}{\partial x_e} + \sum_{j=1}^{M} \psi_j \left[\sum_{k=1}^{M} \left(\frac{\partial K_{jk}}{\partial x_e} u_k + K_{jk} \frac{\partial u_k}{\partial x_e} \right) - \frac{\partial F_j}{\partial x_e} \right]$$

$$= \frac{\partial c}{\partial x_e} + \sum_{k=1}^{M} \left(\frac{\partial c}{\partial u_k} + \sum_{j} \psi_j K_{jk} \right) \frac{\partial u_k}{\partial x_e} + \sum_{j=1}^{M} \psi_j \left(\sum_{k} \frac{\partial K_{jk}}{\partial x_e} u_k - \frac{\partial F_j}{\partial x_e} \right) \tag{3.20}$$

At first glance, the attempt has only led to a more complicated expression. It can be exploited, however, that the factors ψ_j may be chosen arbitrarily without changing the value of the differential. This includes the possibility of computing them in such a way that the term $\frac{\partial u_j}{\partial x_e}$ vanishes. This is the case when the following applies

$$\mathbf{K}\psi = \delta \tag{3.21}$$

where $\delta = \left[-\frac{\partial c}{\partial u_1}, \dots, -\frac{\partial c}{\partial u_M} \right]^{\mathrm{T}}$ is a pseudo-load vector. The similarity of Eq. (3.21) with the basic FEM equation is obvious and it is solved to determine the multipliers ψ_j. In doing so, the total differential simplifies to:

$$\frac{dc\,(\mathbf{x}, \mathbf{U}(\mathbf{x}))}{dx_e} = \frac{\partial c}{\partial x_e} + \sum_{j=1}^{M} \psi_j \left(\sum_{k=1}^{M} \frac{\partial K_{jk}}{\partial x_e} u_k - \frac{\partial F_j}{\partial x_e} \right) \tag{3.22}$$

Additionally, $\frac{\partial F_j}{\partial x_e} = 0$ applies, since the loads are typically defined independently of the design variables. The remaining terms can be determined straightforwardly.

Consequently, the adjoint method solves a system of equations in each iteration to determine the sensitivities. In terms of computational cost, this is quite inexpensive.

3.5 Solution Methods

3.5.1 Mathematical Programming

One way to solve the constrained nonlinear optimization problem is to employ the sensitivity information used within mathematical programming for updating the design variables [8, 10, 17–21]. In many cases, it is useful to locally approximate the objective and constraint functions convexly and thus reduce the complexity of the optimization problem. Such local approximation techniques rely on Taylor series expansion. The design variables are then developed for the approximated subproblem instead of for the original formulation. In doing so, the effort required to solve the problem in the current iteration is reduced considerably. The various approximation approaches differ in the polynomial degree used to locally approximate the objective and constraint functions.

In sequential linear programming (SLP) [6, 20, 30] both the objective as well as the constraint functions are approximated linearly. In contrast, sequential quadratic programming (SQP) [31–38] uses a quadratic approach for the objective and a linear one for the constraint functions. Convex Linearization (CONLIN) [39, 40], on the other hand, uses the sign of the partial derivatives to determine whether the Taylor series expands linearly or reciprocally. Finally, the method of moving asymptotes (MMA) [6, 41, 42] should be mentioned, which is an enhancement of CONLIN, thus more adaptive and very versatile applicable to a broad range of problems.

3.5.2 Optimality Conditions (OC)

Approaches based on OC include the definition of a mathematical representation of a system state that is considered optimal [8, 10, 12, 43]. Based on this, an iteration scheme for updating the design variables is developed, which evolves the system toward the defined optimal state. Consequently, OC-based methods are not universally applicable but have to be developed individually for each optimization problem. However, they are usually very efficient since they are customized.

OC methods can be subdivided into mathematical and empirical approaches. In case of the mathematical approaches, the expression describing the optimal system state is determined from the stationarity condition of the Lagrange function, namely from the KKT conditions. The empirical approaches, on the other hand, rely on observations and empirical assumptions about the optimal state of a structure. Examples are the Fully Stressed Design [12], which, however, only guarantees weight-optimized solutions for statically determined boundary conditions, and the Biological Growth Rule [13, 43, 44] that strives for uniform surface stress, since this is observed in nature for trees, bones, tiger claws, and antlers.

The approaches presented in this book using OC employ the former, that is, mathematical methods.

References

1 Zienkiewicz, O.C., Taylor, R.L., Nithiarasu, P., and Zhu, J.Z. (1977). *The Finite Element Method*, Vol. 3. London: McGraw-Hill.

2 Bathe, K.-J. (1996). *Finite Element Procedures, Prentice Hall International Editions*. Englewood Cliffs, NJ: Prentice Hall. ISBN: 013349697x.

3 Hughes, T.J.R. (2000). *The Finite Element Method: Linear Static and Dynamic Finite Element Analysis*. Mineola, NY: Dover, Reprint. ISBN: 0486411818. http://www.loc.gov/catdir/description/dover031/00038414.html.

4 Maute, K., Schwarz, St., and Ramm, E. (1999). Structural optimization — The interaction between form and mechanics. *Journal of Applied Mathematics and Mechanics* 79 (10): 651–673.

5 Vanderplaats, G.N. (1999). Structural design optimization status and direction. *Journal of Aircraft* 31 (01): 11–20. https://doi.org/10.2514/2.2440.

6 Harzheim, L. (2008). *Strukturoptimierung: Grundlagen und Anwendungen.* Frankfurt am Main: Harri Deutsch.

7 Schumacher, A. (2013). *Optimierung Mechanischer Strukturen: Grundlagen und industrielle Anwendungen.* Berlin, Heidelberg: Springer-Verlag. ISBN: 978-3-642-34699-6.

8 Belegundu, A.D. and Chandrupatla, T.R. (2014). *Optimization Concepts and Applications in Engineering.* New York: Cambridge University Press. ISBN: 0521878462.

9 Bendsøe, M.P. and Sigmund, O. (2004). *Topology Optimization: Theory, Methods, and Applications.* Berlin: Springer-Verlag.

10 Haftka, R.T. and Gürdal, Z. (1993). *Elements of Structural Optimization.* Dordrecht: Kluwer.

11 Xie, Y.M. and Steven, G.P. (1997). *Evolutionary Structural Optimization.* Springer. ISBN: 9781447112501.

12 Baier, H., Seeßelberg, C., and Specht, B. (1994). *Optimierung in der Strukturmechanik.* Braunschweig: Vieweg + Teubner. ISBN: 9783322907004.

13 Mattheck, C. (1990). Engineering Components grow like trees. *Materialwissenschaft und Werkstofftechnik* 21 (4): 143–168. https://doi.org/10.1002/MAWE .19900210403.

14 Rossow, M.P. and Taylor, J.E. (1973). A finite element method for the optimal design of variable thickness sheets. *AIAA Journal* 11: 1566–1569.

15 Mark, P. (2003). Optimisation methods for the bending design of reinforced concrete sections. *Beton- und Stahlbetonbau* 98 (9): 511–519.

16 Mark, P. (2004). Optimisation of foundation areas and calculation of foundation pressures using optimisation methods and spreadsheet analysis. *Bautechnik* 81 (1): 38–43.

17 Fletcher, R. (1987). *Practical Methods of Optimization.* Chichester: Wiley. ISBN: 978-0-471-91547-8.

18 Papageorgiou, M., Leibold, M., and Buss, M. (2015). *Optimierung: Statische, dynamische, stochastische Verfahren für die Anwendung.* Berlin, Heidelberg: Springer Vieweg. ISBN: 9783662469361.

19 Nocedal, J. and Wright, S.J. (2006). *Numerical Optimization.* New York: Springer. ISBN: 9780387400655.

20 Arora, J.S. (2011). *Introduction to Optimum Design.* Amsterdam: Elsevier Academic Press.

21 Bhatti, M.A. (2000). *Practical Optimization Methods.* New York: Springer.

22 Kuhn, H.W. and Tucker, A.W. (1951). Nonlinear programming. *Proceedings of the 2nd Berkeley Symposium* (ed. Jerzy Neyman), pp. 481–492.

23 Lasdon, L.S. (1970). *Optimization Theory for Large Systems.* London: The MacMillan Company.

24 Mangasarian, O.L. (1969). *Nonlinear Programming*. McGraw-Hill.

25 Wolfe, P. (1961). A duality theorem for nonlinear programming. *Quarterly of Applied Mathematics* 19 (3): 239–244. https://doi.org/10.1090/qam/135625.

26 Schwarz, S. (2011). Sensitivity analysis and optimization for nonlinear structural response. PhD thesis. Stuttgart: University of Stuttgart.

27 Thomas, H., Zhou, M., and Schramm, U. (2002). Issues of commercial optimization software development. *Structural and Multidisciplinary Optimization* 23 (2): 97–110. https://doi.org/10.1007/s00158-002-0170-x.

28 Tortorelli, D.A. and Michaleris, P. (1994). Design sensitivity analysis: overview and review. *Inverse Problems in Engineering* 1 (1): 71–105. https://doi.org/10.1080/174159794088027573.

29 Dems, K. (1991). First- and second-order shape sensitivity analysis of structures. *Structural and Multidisciplinary Optimization* 3 (2): 79–88. https://doi.org/10.1007/BF01743276.

30 Papageorgiou, M., Leibold, M., and Buss, M. (2015). *Optimierung. Statische, dynamische, stochastische Verfahren für die Anwendung*. Berlin, Heidelberg: Springer Vieweg.

31 Bartholomew, P. and Biggs, M.C. (1979). An Improved Implementation of the Recursive Quadratic Programming Method for Constrained Minimization. *Technical Report No. 105*. The Hatfield Polytechnic, Hatfield, England: Numerical Optimisation Centre.

32 Biggs, M.C. (1972). Constraint minimization using recursive equality quadratic programming. In: *Numerical Methods for Nonlinear Optimization* (ed. F.A. Lootsma), 411–428. London: Academic Press.

33 Biggs, M.C. (1975). Constraint minimization using recursive quadratic programming: some alternative subproblem formulations. In: *Towards Global Optimization* (eds. L.C.W. Dixon and G.P. Szego), 341–349. North-Holland Publishing Co.

34 Han, S.P. (1977). A globally convergent method for nonlinear programming. *Journal of Optimization Theory and Applications* 22 (3): 297–309. https://doi.org/10.1007/BF00932858.

35 Murray, W. (1976). *Methods for Constraint Optimization. Optimization in Action*. New York: L. C. W. Dixon.

36 Powell, M.J.D. (1978). Algorithms for nonlinear constraints that use lagrangian functions. *Mathematical Programming* 14 (1): 224–248.

37 Powell, M.J.D. (1978). A fast algorithm for nonlinearly constrained optimization calculations. In: *Numerical Analysis, Lecture Notes in Mathematics*, Vol. 630 (ed. G.A. Watson), 144–157. Berlin, Heidelberg: Springer. https://doi.org/10.1007/BFb0067703.

38 Schittkowski, K. (1983). On the convergence of a sequential quadratic programming method with an augmented lagrangian line search function. *Mathematische Operationsforschung und Statistik. Series Optimization* 14 (2): 197–216.

39 Fleury, C. and Braibant, V. (1986). Structural optimization: a new dual method using mixed variables. *International Journal for Numerical Methods in Engineering* 23 (3): 409–428. https://doi.org/10.1002/nme.1620230307.

40 Fleury, C. (1989). First and second order convex approximation strategies in structural optimization. *Structural and Multidisciplinary Optimization* 1 (1): 3–10. https://doi.org/10.1007/BF01743804.

41 Svanberg, K. (1987). The method of moving asymptotes: A new method for structural optimization. *International Journal for Numerical Methods in Engineering* 24 (2): 359–373. https://doi.org/10.1002/nme.1620240207.

42 Svanberg, K. (2014). MMA and GCMMA - two methods for nonlinear optimization. https://people.kth.se/~krille/mmagcmma.pdf (accessed 19 September 2021).

43 Mattheck, C. (1998). *Design in Nature: Learning from Trees*. Berlin, New York: Springer. ISBN: 9783642587474.

44 Mattheck, C. (1990). Design and growth rules for biological structures and their application to engineering. *Fatigue and Fracture of Engineering Materials and Structures* 13 (5): 535–550. https://doi.org/10.1111/j.1460-2695.1990.tb00623.x.

4

Identification of Structures

Key learnings after reading this chapter:

- What are practical approaches for identifying global optimal structure designs?
- What are typical results and how do different optimization parameters influence them?
- How can optimization results be transferred into structure designs?

This chapter discusses optimization methods as a tool for two types of problems (Figure 4.1).

The first is the identification of a good structure design for a given building material within a permissible design space and under consideration of static and geometrical boundary conditions in the draft phase. The principles are three-fold: resource efficiency, cost-effectiveness, and material compatibility. For the latter, it is particularly important to ensure that the designs are adapted to the material-specific properties, namely compression or tension affinity.

The second type of problem is to improve the overall design of standard components such as beams, girders, and walls. Starting from the given basic shape, the aim here is to reduce material consumption without compromising stiffness and load-bearing capacity. For this purpose, the amount of material available to form the structure is first reduced and then rearranged toward the internal load transfer (principal stress trajectories). The results yield lightweight reinforced concrete (RC) and concrete–steel hybrid structure designs.

Optimization Aided Design: Reinforced Concrete, First Edition.
Georgios Gaganelis, Peter Mark, and Patrick Forman.
© 2022 Ernst & Sohn GmbH & Co.KG. Published 2022 by Ernst & Sohn GmbH & Co.KG.

Identification of optimized designs

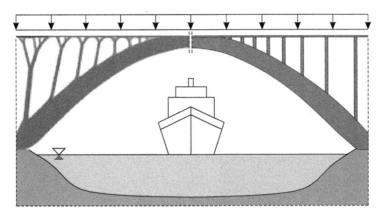

Improvement of conventional designs

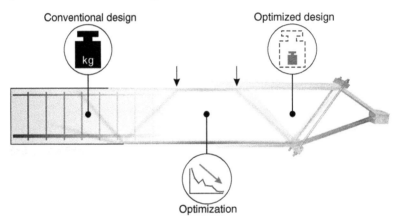

Figure 4.1 Overview of Chapter 4: identification of structures.

4.1 One-material Structures

Related Examples 4.1–4.10.

4.1.1 Problem Statement

In topology optimization, a limited amount of material is iteratively redistributed within a predefined design space in such a way that the objective function is minimized until a convergence criterion is met. The material distribution is oriented to the internal force flow according to the "form follows force" principle. Using continuum finite elements (continuum topology optimization, CTO), topology optimization can be applied to reduce a structure's material consumption to the necessary load-bearing system by eliminating any excess material. In this way, resource efficiency is aimed at already at the (pre-)design stage. This is particularly important for RC structures, since their main components, namely cement, concrete, and reinforcing steel, emit vast quantities of greenhouse gas (GHG) in production. For classification, the construction industry accounts for 25 % of worldwide CO_2 emissions [1], with cement alone accounting for 5–10 % [2–5] as already mentioned in Section 1.3. In this regard, CTO can be used as a design tool for distinctly more sustainable structures [6–8].

Typically, the optimization problem's objective function is defined to minimize the mean structural compliance (c), which equals stiffness maximization. In discrete finite element (FE) notation, the compliance reads

$$c = \mathbf{FU} = \mathbf{U}^{\mathsf{T}}\mathbf{KU} \tag{4.1}$$

where $\mathbf{F}$, $\mathbf{U}$, and $\mathbf{K}$ are the global load vector, displacement vector, and stiffness matrix, respectively. The available material volume (V), which is limited to a fraction $\beta \in [0, 1]$ of the initial volume (V^0), serves as constraint in order to prevent trivial solutions in which the design space is completely filled with material to achieve maximum stiffness.

The FE analysis model is linked to the optimization problem via material interpolation schemes [9]. The most popular is SIMP (Solid Isotropic Material with Penalization) [9–14]. In SIMP, each element e is assigned a design variable $x_e \in [0, 1]$, where the latter can be considered normalized element densities:

$$x_e = \frac{\rho_e}{\rho_e^0} \tag{4.2}$$

In Eq. (4.2), ρ_e is the associated element density and ρ_e^0 is the physical density of the employed material. Through stiffness adjustment of each element, the design variables define also their material assignment:

$$E_e(x_e) = x_e^p E^0 \tag{4.3}$$

where E_e and E^0 are the elemental and physical Young's modulus, respectively. The penalty exponent $p > 1$ gives preference to a 0-1 distribution of the design variables by assigning underproportional stiffness to intermediate values of the design variables (Figure 4.2). In this way, they become inefficient for minimizing the structural compliance, hence they are avoided by the optimization algorithm. This is reasonable, because intermediate densities are difficult to interpret in practice [15]. A reasonable value for p can be justified physically with respect to the Poisson's ratio of the employed material [9, 11]. However, generally it suffices to simply set $p = 3$.

The optimization problem can then be expressed comprehensively as follows:

$$
\begin{aligned}
\text{find:} \quad & \mathbf{x} = [x_1, x_2, \ldots, x_{N_e}]^{\mathsf{T}} \\
\text{such that:} \quad & f(\mathbf{x}) = c = \mathbf{U}^{\mathsf{T}}\mathbf{K}(\mathbf{x})\mathbf{U} = \sum_{e=1}^{N_e} \mathbf{u}_e^{\mathsf{T}} \mathbf{k}_e(x_e)\mathbf{u}_e \to \min_{\mathbf{x}} \\
\text{subject to:} \quad & g(\mathbf{x}) = V - \beta V^0 = \sum_{e=1}^{N_e} x_e v_e - \beta \sum_{e=1}^{N_e} v_e \leq 0 \\
& 0 < x_e^L \leq x_e \leq x_e^U \qquad\qquad e \in [1, N_e]
\end{aligned}
\tag{4.4}
$$

In Eq. (4.4), N_e is the number of elements, $\mathbf{k}_e$ and $\mathbf{u}_e$ are the element's stiffness matrix and displacement vector, respectively, v_e is the volume of element e, $x_e^L = 10^{-3}$ is a

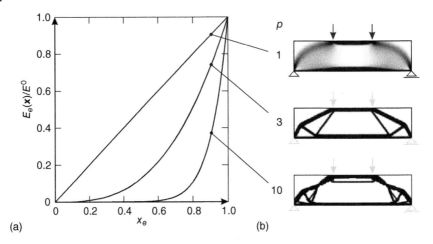

Figure 4.2 Penalty exponent p in SIMP: (a) stiffness assignment, (b) impact on results.

lower nonzero bound of the design variables to avoid singularity and $x_e^U = 1$ is the upper bound, which represents the physical density.

4.1.2 Sensitivity Analysis

The partial derivatives of both the objective and the constraint function, the so-called sensitivities, are required to solve the optimization problem and update the design variables. They can be derived following the adjoint method. For further details, the reader is referred to, for instance, Bendsøe and Sigmund [11]. In doing so, the adjoint method approach leads to:

$$\frac{\partial f(\mathbf{x})}{\partial x_e} = -\mathbf{U}^\mathsf{T} \frac{\partial \mathbf{K}(\mathbf{x})}{\partial x_e} \mathbf{U} = -px_e^{(p-1)} \mathbf{u}_e^\mathsf{T} \mathbf{k}_e \mathbf{u}_e \tag{4.5}$$

for the objective function sensitivities. The sensitivities of the constraint function, on the other hand, are computed straightforwardly to simply:

$$\frac{\partial g(\mathbf{x})}{\partial x_e} = v_e \tag{4.6}$$

4.1.3 Filtering

Two well-known numerical problems are encountered in CTO [11, 15, 16]. The first is the so-called checkerboard problem. These checkerboard patterns refer to elements containing material, which are connected at their vertices (hinged elements) and cause an oscillating material distribution as shown in Figure 4.3a. The reason for this lies in a numerically overestimated stiffness of such hinged elements when using linear shape functions for approximation in the FE model.

The second problem is known as mesh dependency of the optimization results. If the number of elements forming the FE mesh increases, the resultant structures show more delicate struts and smaller holes as they exhibit higher stiffness values

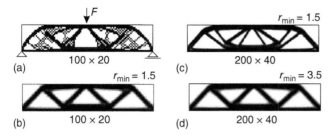

Figure 4.3 Numerical problems in topology optimization: (a) checkerboards, (b, c) mesh dependency, (d) mesh independent filtering.

and are therefore preferred by the optimization algorithm. Thus, instead of a finer FE mesh providing a more accurate numerical approximation of the mechanical behavior, it instead affects the optimization result which is highly undesired. This effect is demonstrated in Figure 4.3b–c where the optimization results for a design space of 100 × 20 and 200 × 40 elements, respectively, are compared.

Various techniques have been developed to overcome these problems [11, 15, 17]. The most popular are the so-called mesh independent filters, inspired by image processing [16]. From a mathematical point of view, by using filters within the optimization, the permissible solution space is limited to a subspace through the applied filter radius r_{min}. From an engineering point of view, however, r_{min} defines a lower limit for the permissible thickness of structural components forming the resulting structure, cf. Figure 4.3d. The filter radius can therefore be interpreted as a kind of a manufacturing constraint.

Among a multitude of others, the most important filters are the so-called density and the sensitivity filter [11, 15]. For the sake of simplicity, the following explanations are limited to the sensitivity filter [18] since it is easy to implement and sufficient for practical application. It modifies the sensitivities of the objective function at the end of an iteration, right before the design variable update as can be seen from the flow chart in Figure 4.4. The modified sensitivities of an element are then calculated as the mean value of all adjacent elements' sensitivities, lying within r_{min} and weighted by their distance. Thus, the modified objective function sensitivity of an element e is

$$\frac{\partial \tilde{f}}{\partial x_e} = \frac{1}{x_e \sum_{i=1}^{N_e} H_{ei}} \sum_{i=1}^{N_e} H_{ei} x_i \frac{\partial f}{\partial x_i} \tag{4.7}$$

where

$$H_{ei} = \begin{cases} r_{min} - \Delta_{ei} & \text{if } \Delta_{ei} \leq r_{min} \\ 0 & \text{else} \end{cases} \tag{4.8}$$

with Δ_{ei} being the distance between the center of element i and e.

The sensitivity filter is easy to implement since the optimization problem remains unchanged. It should be emphasized that it works heuristically and is not derived mathematically well founded. For practical applications, however, the advantage of simplicity prevails.

4.1.4 Solving

The optimization problem can be solved using a nonlinear optimization algorithm or, alternatively, by applying a heuristic update scheme for the design variables, which is developed using an optimality criterion (OC) derived from the Lagrangian function. The latter is a customized approach to the problem, which has the advantage of being numerically robust and converging to a solution rapidly [6, 19]. For the sake of efficient practical application, the OC-based update scheme is described in more detail below.

The Lagrangian function of the optimization problem stated in Eq. (4.4) reads .

$$L = \mathbf{U}^\mathsf{T}\mathbf{K}\mathbf{U} + \Lambda\left(V - \beta V^0\right) + \sum_{e=1}^{N_e} \lambda_e\left(x_e^L - x_e\right) + \sum_{e=1}^{N_e} \gamma_e\left(x_e - x_e^U\right) \tag{4.9}$$

where $\Lambda \geq 0$, $\lambda \geq 0$ and $\gamma_e \geq 0$ are the Lagrangian multipliers. From the stationarity condition with respect to the design variables follows:

$$\frac{\partial L}{\partial x_e} = \left(\frac{\partial \mathbf{U}^\mathsf{T}}{\partial x_e}\mathbf{K}\mathbf{U} + \mathbf{U}^\mathsf{T}\frac{\partial \mathbf{K}}{\partial x_e} + \mathbf{U}^\mathsf{T}\mathbf{K}\frac{\partial \mathbf{U}}{\partial x_e}\right) + \Lambda\frac{\partial g}{\partial x_e} - \lambda_e + \gamma_e = 0 \tag{4.10}$$

Assuming that the loads are independent of the design variables, it can be shown [11] that

$$\mathbf{U}^\mathsf{T}\mathbf{K}\frac{\partial \mathbf{U}}{\partial x_e} = -\mathbf{U}^\mathsf{T}\frac{\partial \mathbf{K}}{\partial x_e}\mathbf{U} \tag{4.11}$$

as well as

$$\mathbf{U}^\mathsf{T}\mathbf{K}\frac{\partial \mathbf{U}}{\partial x_e} = \frac{\partial \mathbf{U}^\mathsf{T}}{\partial x_e}\mathbf{K}\mathbf{U} \tag{4.12}$$

holds true. Substituting Eqs. (4.11) and (4.12) in (4.10) leads to the reformulated stationarity condition:

$$\frac{\partial L}{\partial x_e} = \underbrace{-\mathbf{U}^\mathsf{T}\frac{\partial \mathbf{K}}{\partial x_e}\mathbf{U} + \Lambda\frac{\partial g}{\partial x_e}}_{\frac{\partial f}{\partial x_e}} - \lambda_e + \gamma_e = 0 \tag{4.13}$$

The OC is then obtained by rearranging Eq. (4.13) which finally yields

$$\underbrace{\frac{-\frac{\partial f}{\partial x_e}}{\Lambda\frac{\partial g}{\partial x_e}}}_{G_e} = 1 - \frac{\lambda_e}{\Lambda v_e} + \frac{\gamma_e}{\Lambda v_e} \tag{4.14}$$

By evaluating Eq. (4.34), an update scheme for the design variables can be derived as described in [11, 12, 20, 21], such that G_e approaches 1. Such an update scheme might read, for instance, as follows:

$$x_e^{(k+1)} = \begin{cases} M_x^- & \text{if} \quad \left[G_e^{(k)}\right]^{0.5} x_e^{(k)} \leq M_x^- \\ M_x^+ & \text{if} \quad \left[G_e^{(k)}\right]^{0.5} x_e^{(k)} \geq M_x^+ \\ \left[G_e^{(k)}\right]^{0.5} x_e^{(k)} & \text{else} \end{cases} \tag{4.15}$$

where

$$M_x^- = \max \left\{ \left(1 - \mu_x\right) x_e^{(k)}, x_e^L \right\} \tag{4.16a}$$

$$M_x^+ = \min \left\{ x_e^U, \left(1 + \mu_x\right) x_e^{(k)} \right\} \tag{4.16b}$$

Here, x_e^L and x_e^U are the lower and upper bound of the design variables, μ_x is a move limit, which prevents too large changes of the design variables between iterations to avoid convergence problems and k denotes the current iteration number. Appropriate values for the parameters are, for example, $x_e^L = 10^{-3}$, $x_e^U = 1$, and $\mu_x = 0.2$. The Lagrange multiplier Λ in Eq. (4.34) is computed numerically using a bisection algorithm in such a way that the volume constraint is met, since it can be expected that the stiffest structure is obtained by exploiting the maximum amount of permissible material.

4.1.5 Optimization Process

Figure 4.4 shows the flow chart of the one-material topology optimization approach. First, the model is initialized, i.e. the design space, boundary conditions, material properties, and optimization parameters are defined. Then, the first FE analysis is conducted and the corresponding displacement field is computed. From this, the principal stresses and strains are then determined. The associated sensitivities of the objective and constraint functions are then evaluated. In order to counteract well-known numerical problems in optimization with linear elastic FE models, the former are filtered. The sensitivity information of both objective and constraint

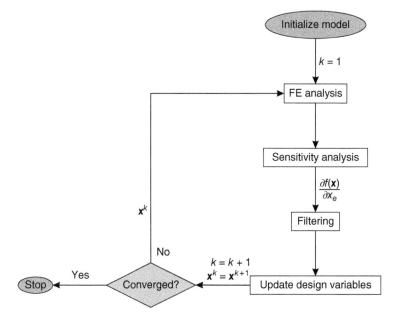

Figure 4.4 Flow chart of the topology optimization approach.

function are then used by the optimizer, for instance, the introduced OC-based update scheme, to update the design variables. Finally, at the end of the iteration, a convergence criterion is checked. A suitable one could, for example, be the largest difference between the design variables of two subsequent iterations:

$$\max \Delta x_e = \max \left| x_e^{(k+1)} - x_e^{(k)} \right| \leq \text{tol}_x \tag{4.17}$$

with $\text{tol}_x = 10^{-2}$ being a reasonable limit. If the convergence criterion is met, the optimization is terminated. If not, a new loop is initiated by performing a linear elastic FE analysis.

4.1.6 Multiple Load Cases

Usually, it is necessary to consider more than one load case when determining internal forces and designing structural components. The same applies to the optimization of the structural design, where the existing load cases must also be adequately accounted for. The optimized material distribution differs significantly depending on whether all loads are considered within one or within individual load cases. The single load case results are often unstable designs consisting of rectangular substructures, whereas a multiple load case approach yields stable structures composed of triangular segments, cf. Figure 4.5. The two forces F work like dead loads and act at the same time (a). On the contrary, F_1 and F_2 may occur individually from each other, so both loads or just one of them might be present in a sense of live loads (b).

In the case of the standard compliance minimization approach (Eq. (4.4)), multiple load cases can be taken into account by reformulating the objective function as minimizing the weighted sum of the compliance values obtained from all individual load cases [11]:

$$f(\mathbf{x}) = \sum_{l=1}^{M} w_l c_l = \sum_{l=1}^{M} w_l \mathbf{U}_l^\top \mathbf{K}(\mathbf{x}) \mathbf{U}_l \rightarrow \min_x \tag{4.18}$$

Here, l represents the load case number, M is the total number of all load cases, w_l are the corresponding weighting factors, c_l are the compliance values, $\mathbf{U}_l$ are the respective displacement vectors, and $\mathbf{K}(\mathbf{x})$ is the common global stiffness matrix. The displacement vectors are computed independently from each other, from which then the compliance of each corresponding load case is determined. In turn,

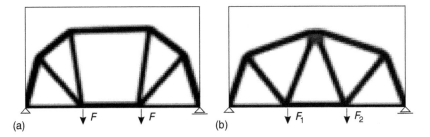

(a) (b)

Figure 4.5 Different optimization results depending on the load case definition: (a) one load case, (b) two load cases.

the compliance values are then multiplied by the respective weighting factor, if necessary, and summed up to the total objective function value. Its sensitivities are determined in the same way as weighted sum:

$$\frac{\partial f(\mathbf{x})}{\partial x_e} = \sum_{l=1}^{M} w_l \left(-\mathbf{U}_l^T \frac{\partial \mathbf{K}(\mathbf{x})}{\partial x_e} \mathbf{U}_l \right) \tag{4.19}$$

The multiple load case problem can be solved with common nonlinear optimization algorithms as well as with the OC-based update scheme given in Eq. (4.15) by substituting Eqs. (4.18) and (4.19) into the objective function and its sensitivities.

4.2 One-material Stress-biased Structures

Related Example: 4.11.

4.2.1 Problem Statement

The main building materials, concrete and steel, show different mechanical properties. Concrete, on the one hand, exhibits a pronounced compressive strength but negligible tensile load-bearing capacities. Due to its very low price and world-wide availability, however, it is the dominant building material and is reinforced by steel reinforcement in structural areas subject to tensile stresses. Steel, on the other hand, has good load-bearing properties in both compression and tension. But because of its high costs, its use is rather reasonable for components predom-inantly exposed to tensile loading. The global structural design should therefore be material-dependent, i.e. compression-dominant when using concrete and tension-dominant when using steel. Therefore, topology optimization approaches are needed that produce compression- or tension-dominant structural design proposals.

The optimization problem is set up as classical compliance minimization with vol-ume constraint using continuum elements:

$$\begin{aligned}
\text{find:} \quad & \mathbf{x} = [x_1, x_2, \dots, x_{N_e}]^T \\
\text{such that:} \quad & f(\mathbf{x}) = c = \mathbf{U}^T \mathbf{K}(\mathbf{x})\mathbf{U} = \sum_{e=1}^{N_e} \mathbf{u}_e^T \mathbf{k}_e(x_e)\mathbf{u}_e \rightarrow \min_{\mathbf{x}} \\
\text{subject to:} \quad & g(\mathbf{x}) = V - \beta V^0 = \sum_{e=1}^{N_e} x_e v_e - \beta \sum_{e=1}^{N_e} v_e \leq 0 \\
& 0 \leq x_e \leq 1 \qquad\qquad\qquad\qquad\qquad\qquad\quad e \in [1, N_e]
\end{aligned} \tag{4.20}$$

Here, $\mathbf{x}$ is the vector containing all N_e design variables, each representing a rela-tive element density, where $x_e = 0$ indicates a void and $x_e = 1$ a solid element with full material allocation, c is the mean structural compliance (reciprocal of structural stiffness), $\mathbf{U}$ and $\mathbf{u}_e$ are the global and local displacement vectors, respectively, V is the structural volume and V^0 is the design space's initial volume, where $\beta \in [0,1]$ represents their ratio and defines the targeted residual volume for the optimized structure, v_e is the volume of element e, $\mathbf{K}$ and $\mathbf{k}_e$ are the global and local stiffness matrices, respectively. The latter is interpolated element-wise via a slightly modified SIMP approach [9] with respect to the design variables and reads

$$\mathbf{k}_e(x_e) = \left[E_{\min} + x_e^p \left(E^0 - E_{\min} \right) \right] \mathbf{k}_e^0 \tag{4.21}$$

where $E_{\min} = 10^{-6}$, for instance, is a lower nonzero bound of the Young's modulus to circumvent singular stiffness matrices in case where $x_e = 0$, E^0 is the physical Young's modulus of the employed material, p is a penalty exponent to achieve a quasi 0-1 distribution of the design variables which can generally be set to $p = 3$, and $\mathbf{k}_e^0$ is the element's stiffness matrix independent from the Young's modulus.

The optimization problem within a linear elastic FE framework cannot take into account a material-related compression or tension affinity. However, stress-specific steering of the optimization process is approached in an indirect way by systematically modifying the sensitivities [22]. Details will be presented subsequently.

4.2.2 Sensitivity Analysis and Stress Bias

The derivation of stress bias starts from the objective function sensitivities of the basic notation given in Eq. (4.20) using the adjoint method [21, 23]:

$$\frac{\partial f}{\partial x_e} = -px_e^{(p-1)} \left(E^0 - E_{\min} \right) \mathbf{u}_e^{\mathsf{T}} \mathbf{k}_e^0 \mathbf{u}_e \tag{4.22}$$

The sensitivities of the volume constraint can be computed directly and simply read

$$\frac{\partial g}{\partial x_e} = v_e \tag{4.23}$$

For an easy heuristic control of the material distribution toward compressive or tensile structure designs, the sensitivities of the constraint function are transformed to "weighted element volumes" through modification factors, which are determined depending on the respective element's stress state s_e:

$$\left. \frac{\partial g}{\partial x_e} \right|_{\mathrm{mod}} = v_e \psi_e(s_e) \tag{4.24}$$

The optimization problem is, in other words, extended by a penalty term, which, however, is only used for altering the constraint sensitivities. In Eq. (4.24), ψ_e is a stress state-dependent weighting factor, which is computed for each element separately as follows:

$$\psi_e(s_e) = \begin{cases} \Psi^+ & \text{for } s_e > 0 \quad \text{(pure tensile stress state)} \\ \Psi_e(s_e) & \text{for } s_e \leq 0 \quad \text{(mixed stress state)} \\ \Psi^- & \text{for } s_e > 0 \quad \text{(pure compressive stress state)} \end{cases} \tag{4.25}$$

Here, Ψ^+ and Ψ^- are the weighting factors for tension-affine and compression-affine material, respectively, whereby it holds that the higher a weighting factor value is, the higher the respective stress state is penalized, i.e.:

$$\begin{aligned} \Psi^+/\Psi^- &> 1 \quad \text{compression-affine} \\ \Psi^+/\Psi^- &= 1 \quad \text{unbiased} \\ \Psi^+/\Psi^- &< 1 \quad \text{tension-affine} \end{aligned} \tag{4.26}$$

The stress ratio s_e of an element is defined in plane stress condition as:

$$s_e = \frac{\sigma_{|\min|} \; \text{sgn}(\sigma_{1,e})}{\sigma_{|\max|} \; \text{sgn}(\sigma_{2,e})} \tag{4.27}$$

with

$$
\begin{aligned}
\sigma_{|\min|} &= \min\{|\sigma_{1,e}|, |\sigma_{2,e}|\} \\
\sigma_{|\max|} &= \max\{|\sigma_{1,e}|, |\sigma_{2,e}|\}
\end{aligned} \tag{4.28}
$$

being the absolute max and min values of the element principal stresses $\sigma_{1,e}$ and $\sigma_{2,e}$. The sgn function takes the sign of the stress values into account and is generally defined as:

$$
\text{sgn}(\sigma_{i,e}) = \begin{cases} -1 & \text{if } \sigma_{i,e} < 0 \\ 0 & \text{if } \sigma_{i,e} = 0 \\ 1 & \text{if } \sigma_{i,e} > 0 \end{cases} \tag{4.29}
$$

In case $s_e > 0$, the element exhibits stresses of only one sign, i.e. it is either purely under compression or purely under tension. Therefore, assigning a weighting factor is straightforward: for a pure tensile stress state, it holds that $\sigma_{1,e}, \sigma_{2,e} > 0$, and thus $\psi_e(s_e) = \Psi^+$, whereas for a pure compressive stress state the opposite applies, namely $\sigma_{1,e}, \sigma_{2,e} < 0$ and $\psi_e(s_e) = \Psi^-$ (cf. Eq. (4.25)). On the other hand, in case $s_e < 0$, the element shows stresses of both signs, which is undesirable, but can occur particularly in the first iterations and in elements forming structural nodes. If this is the case, Ψ_e is interpolated linearly (Figure 4.6) as follows:

$$
\Psi_e(s_e) = \begin{cases} \Psi_{\max} + s_e\left(\Psi_{\max} - \overline{\Psi}\right) & \text{ascending} \\ \Psi_{\min} + s_e\left(\Psi_{\min} - \overline{\Psi}\right) & \text{descending} \end{cases} \tag{4.30}
$$

where

$$
\begin{aligned}
\Psi_{\max} &= \max\left(\Psi^-, \Psi^+\right) \\
\Psi_{\min} &= \min\left(\Psi^-, \Psi^+\right) \\
\overline{\Psi} &= \tfrac{1}{2}\left(\Psi^- + \Psi^+\right)
\end{aligned} \tag{4.31}
$$

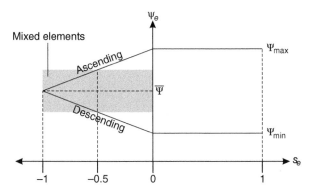

Figure 4.6 Weighting factors with respect to the stress ratio and definition of mixed elements in compression-/tension-biased topology optimization [22].

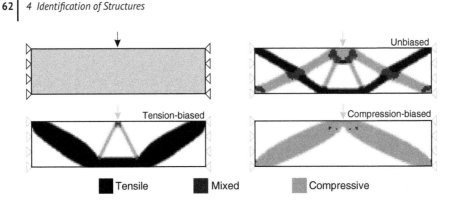

Tension-biased

Compression-biased

■ Tensile　　■ Mixed　　▨ Compressive

Figure 4.7 Unbiased, tension-biased, and compression-biased topology designs [22].

Distinction is made between the ascending and the descending branch. The choice of which of the two is to be used depends on the absolute values of the principal stresses in comparison with each other. If the compressive stress component exceeds the tensile stress, the function is followed which leads to Ψ^-. Otherwise, if tension prevails, the branch that leads to Ψ^+ is used. In the special case that both principal stresses have exactly the same absolute but different signs, $s_e = -1$ is valid and the mean value ($\overline{\Psi}$) is taken as weighting factor. The more the s_e approaches zero, the more whether compression or tension prevails and the weighting factor approaches either Ψ^- or Ψ^+. Thus, depending on the ratio Ψ^-/Ψ^+, the corresponding element's constrain function sensitivity is modified by a favorable or unfavorable stiffness-to-volume ratio. This simple strategy allows for steering optimization results toward compression-affine or tension-affine structural behavior, which differ significantly from the unbiased design proposals as exemplary shown in Figure 4.7. For a better evaluation of the results unwanted mixed elements can be identified, where the different absolute stress values are too similar. As such undesired elements, all those can be defined for which $-1 \leq s_e \leq -0.5$ applies as shown in Figure 4.6.

4.2.3 Solving

The steered topology optimization problem can be solved analogously to the unbiased approach either by means of a nonlinear optimization algorithm or, more efficiently, via the heuristic OC-based update scheme already introduced in Section 4.1.4:

$$x_e^{(k+1)} = \begin{cases} M_x^- & \text{if} & \left[G_e^{(k)}\right]^{0.5} x_e^{(k)} \leq M_x^- \\ M_x^+ & \text{if} & \left[G_e^{(k)}\right]^{0.5} x_e^{(k)} \geq M_x^+ \\ \left[G_e^{(k)}\right]^{0.5} x_e^{(k)} & \text{else} \end{cases} \tag{4.32}$$

where

$$M_x^- = \max\left\{(1-\mu_x)\,x_e^{(k)}, 0\right\} \tag{4.33a}$$

$$M_x^+ = \min\left\{1, (1+\mu_x)\,x_e^{(k)}\right\} \tag{4.33b}$$

and

$$G_e = \frac{-\frac{\partial f}{\partial x_e}}{\Lambda \frac{\partial g}{\partial x_e}\Big|_{mod}} = 1 - \frac{\lambda_e}{\Delta v_e \psi_e} + \frac{\gamma_e}{\Delta v_e \psi_e} \tag{4.34}$$

Here, the sensitivities of the volume constraint need to be replaced by the modified ones according to Eq. (4.24). For regularization, the filter methods presented in Section 4.1.3 are also in this case suitable to circumvent checkerboards and mesh dependency.

4.2.4 Optimization Process

Figure 4.8 shows a flow chart of the stress-biased one-material topology approach. After initializing both analysis and optimization model, the first iteration is initiated via a FE analysis. From the resulting displacement field, the principal stresses are computed, which are employed to determine the weighting factors for each element. The weighting factors are then used to modify the constraint sensitivities. After computing also the objective function sensitivities, they are filtered and applied to update the design variables via the OC-based update scheme. Finally, the convergence criterion is checked examined. If it is met, the optimization is finished, otherwise a subsequent iteration step is initialized. As in the case of the basic one-material topology optimization approach, a possible convergence criterion is, for example, the largest change of a design variable between two iterations:

$$\max \Delta x_e = \max \left| x_e^{(k+1)} - x_e^{(k)} \right| \le \mathrm{tol}_x \tag{4.35}$$

where $\mathrm{tol}_x = 10^{-2}$ is a reasonable threshold, too.

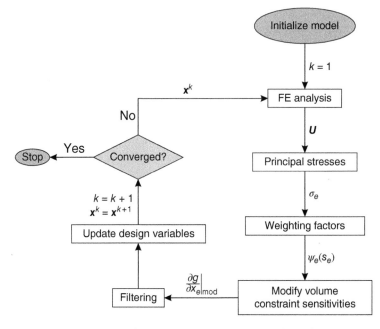

Figure 4.8 Flow chart of the stress-biased topology optimization approach.

4.3 Bi-material Structures

Related Examples: 4.12–4.16.

The two most commonly used construction materials are concrete and steel. They both have their respective benefits and shortcomings. Concrete, on the one hand, is rather cheap, almost everywhere available, with high compressive strength but only negligible tensile strength. Steel, on the other hand, exhibits almost equal compressive and tensile capacities, but is very expensive compared to concrete, corrosive and may suffer from uncertain availability. Therefore, efficient engineering designs demand combinations of different materials which complement each other symbiotically [6].

In contrast, classical methods of topology optimization are capable of taking only one material into account. As stated above, however, it is desirable to identify optimized material arrangements of several materials, taking into account their respective mechanical properties, with, for instance, concrete and steel being applied in compression and tension-dominant sections, respectively. For this purpose, it is necessary to extend the original topology optimization problem in such a way that, on the one hand, several materials are taken into account and, on the other hand, the material's respective stress affinity is considered within optimization [24].

4.3.1 Problem Statement

First, the optimization problem is enhanced to two materials. To this end, the bi-material SIMP approach [25, 26] is used:

$$\mathbf{k}_e(\chi, \varphi) = \chi_e^p \left[\varphi_e \mathbf{k}_e^c + (1 - \varphi_e) \, \mathbf{k}_e^s \right] \tag{4.36}$$

where $\mathbf{k}_e$ is the stiffness matrix of element e and $\mathbf{k}_e^c$ and $\mathbf{k}_e^s$ are the local stiffness matrices representing the mechanical properties of concrete and steel, respectively. Two design variables per element now characterize the optimization problem. The first, $\chi_e \in [10^{-3}, 1]$, represents the relative element density and takes values between 10^{-3} (void) and 1 (solid). The lower bound must be nonzero in order to prevent singular stiffness matrices. However, 10^{-3} is to be understood as a recommendation and may well be chosen differently. The design variable also corresponds to x_e in the original single-material SIMP approach, whereas the penalty exponent $p = 3$ pushes its distribution toward (quasi) 0-1 solutions, thus allowing easy physical interpretation of the results. The second design variable, $\varphi \in [0, 1]$, controls the material assignment of an element. Here, $\varphi_e = 0$ refers to steel and $\varphi_e = 1$ corresponds to concrete, with intermediate values resulting in a linear interpolation between the stiffness matrices of both materials. With respect to a practical application, on the one hand, and in order to avoid complex and time-consuming computation when employing nonlinear material behavior [27], on the other hand, the stiffness matrices are defined linear-elastically and thus they only differ in terms of the Young's modulus and the Poisson's ratio from each other.

The optimization problem is formulated as a compliance minimization with volume constraint:

$$\text{find:} \quad \chi = \left[\chi_1, \chi_2, \ldots, \chi_{N_e}\right]^{\mathrm{T}}, \varphi = \left[\varphi_1, \varphi_2, \ldots, \varphi_{N_e}\right]^{\mathrm{T}}$$

$$\text{such that:} \quad f(\chi, \varphi) = c = \mathbf{U}^{\mathrm{T}}\mathbf{K}(\chi, \varphi)\mathbf{U} = \sum_{e=1}^{N_e} \mathbf{u}_e^{\mathrm{T}}\mathbf{k}_e(\chi_e, \varphi_e)\mathbf{u}_e \rightarrow \min_{\chi, \varphi}$$

$$\text{subject to:} \quad g(\chi) = V - \beta V^0 = \sum_{e=1}^{N_e} \chi_e v_e - \beta \sum_{e=1}^{N_e} v_e \le 0$$

$$10^{-3} \le \chi_e \le 1 \qquad\qquad e \in [1, N_e]$$

$$0 \le \varphi_e \le 1 \qquad\qquad e \in [1, N_e]$$

$$(4.37)$$

Here, χ and φ are the vectors with each containing N_e design variables, c is the mean structural compliance, $\mathbf{u}_e$ and $\mathbf{U}$ are the local and global displacement vectors, respectively, $\mathbf{K}$ and $\mathbf{k}_e$ are the global and local stiffness matrices respectively, v_e, V^0, and V are the element's, initial, and structural volume and $\beta \in [0, 1]$ is a volume fraction which has to be predefined to specify the constraint.

Since the stiffness matrices adopt linear elastic material behavior, solving the optimization problem would favor the stiffest material regardless of the element's stress condition to achieve a structure with minimum compliance. To nevertheless allow for a simple compression–tension differentiation of the materials within a linear FE framework, the so-called bi-material MRM (material replacement method) [6, 24, 28] is applied. It allows to abandon both complex nonlinear material models and numerically challenging stress constraints (singularity problem, cf. [29, 30]). Instead, stress differentiation is addressed in an indirect manner by specifically modifying the sensitivities of the objective function that are used to update the design variables.

4.3.2 Sensitivity Analysis and Stress Differentiation

First, the unmodified sensitivities of the problem stated in Eq. (4.37) are required. They are determined, for instance, by applying the adjoint method [21, 23] which leads to

$$\frac{\partial f(\chi, \varphi)}{\partial \chi_e} = -p\chi_e^{(p-1)}\mathbf{u}_e^{\mathrm{T}}\left[\varphi_e\mathbf{k}_e^c + \left(1 - \varphi_e\right)\mathbf{k}_e^s\right]\mathbf{u}_e \qquad (4.38a)$$

$$\frac{\partial f(\chi, \varphi)}{\partial \varphi_e} = \chi_e^p\mathbf{u}_e^{\mathrm{T}}\left[\mathbf{k}_e^c - \mathbf{k}_e^s\right]\mathbf{u}_e \qquad (4.38b)$$

The sensitivities of the constraint function simply

$$\frac{\partial g(\chi)}{\partial \chi_e} = v_e \qquad (4.39a)$$

$$\frac{\partial g(\chi)}{\partial \varphi_e} = 0 \qquad (4.39b)$$

Using modification factors, R_e^c and R_e^s, for the concrete and the steel stiffnesses, respectively, the sensitivities in Eq. (4.38) are then adjusted depending on the stress state of each element:

$$\left.\frac{\partial f(\chi, \varphi)}{\partial \chi_e}\right|_{\text{mod}} = -p\chi_e^{(p-1)} \mathbf{u}_e^\mathsf{T} \left[\varphi_e R_e^c \mathbf{k}_e^c + (1-\varphi_e) R_e^s \mathbf{k}_e^s\right] \mathbf{u}_e \tag{4.40a}$$

$$\left.\frac{\partial f(\chi, \varphi)}{\partial \varphi_e}\right|_{\text{mod}} = \chi_e^p \mathbf{u}_e^\mathsf{T} \left[R_e^c \mathbf{k}_e^c - R_e^s \mathbf{k}_e^s\right] \mathbf{u}_e \tag{4.40b}$$

In order to determine the modification factors, the elastic strain energy of each element is first divided into a compressive and a tensile portion. The respective portion is then set in relation to the total strain energy. For concrete, this is

$$R_e^c = \frac{\sigma_e^- \varepsilon_e^-}{\sigma_e \varepsilon_e} \tag{4.41}$$

whereas for steel it reads

$$R_e^s = \frac{\sigma_e^+ \varepsilon_e^+}{\sigma_e \varepsilon_e} \tag{4.42}$$

Here, σ_e and ε_e are the vectors containing the principal stresses and strains, whereas σ_e^-, ε_e^- and σ_e^+ and ε_e^+ are the vectors containing the stresses and strains which are negative (compression) and nonnegative (tension), respectively. Obviously, it holds that

$$\sigma_e = \sigma_e^- + \sigma_e^+ \tag{4.43}$$

as well as

$$\varepsilon_e = \varepsilon_e^- + \varepsilon_e^+ \tag{4.44}$$

and thus the modification factors are also complementary, namely

$$R_e^c + R_e^s = 1 \tag{4.45}$$

By modifying the stiffness terms in the sensitivities, a compression-only and tension-only material behavior for concrete and steel, respectively, is approximated for updating the design variables. Figure 4.9 shows in (a) a linear elastic behavior

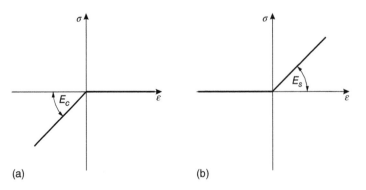

(a) (b)

Figure 4.9 Stress–strain relationship for (a) compression-only, (b) tension-only material.

under compressive strains ($\varepsilon < 0$) and no bearing abilities for tensile strains ($\varepsilon \geq 0$). In (b) this effect is switched with bearing for $\varepsilon > 0$ and $\sigma = 0$ with $\varepsilon \leq 0$.

The elasticity matrix that is needed for calculating the stresses and strains is inter-polated linearly between the elasticity matrix for concrete (D^c) and the one for steel (D^s) in a SIMP-like way:

$$\mathbf{D}_e = \chi_e^p \left[\varphi_e \mathbf{D}^c + (1 - \varphi_e) \mathbf{D}^s \right] \tag{4.46}$$

To give an example, assume an element with the following plane principal stress state:

$$\sigma_e = \begin{bmatrix} -10 \\ 2 \end{bmatrix} \text{N mm}^{-2}, \quad \varepsilon_e = \begin{bmatrix} -0.25 \\ 0.05 \end{bmatrix} \text{\textperthousand}$$

From the spectral decomposition, it follows for the compressive portion,

$$\sigma_e^- = \begin{bmatrix} -10 \\ 0 \end{bmatrix} \text{N mm}^{-2}, \quad \varepsilon_e^- = \begin{bmatrix} -0.25 \\ 0 \end{bmatrix} \text{\textperthousand}$$

and for the tensile portion,

$$\sigma_e^+ = \begin{bmatrix} 0 \\ 2 \end{bmatrix} \text{N mm}^{-2}, \quad \varepsilon_e^+ = \begin{bmatrix} 0 \\ 0.05 \end{bmatrix} \text{\textperthousand}$$

On this basis, the modification factors can be calculated to

$$R_e^c = \frac{(-10) \cdot (-0.25\text{\textperthousand})}{(-10) \cdot (-0.25\text{\textperthousand}) + 2 \cdot 0.05\text{\textperthousand}} = 0.96$$

$$R_e^s = \frac{2 \cdot (0.05\text{\textperthousand})}{(-10) \cdot (-0.25\text{\textperthousand}) + 2 \cdot 0.05\text{\textperthousand}} = 0.04$$

Obviously, the stress state is compression-dominant; hence, the element stiffness matrix of concrete will strongly prevail in the modified sensitivities, since $R_e^c \gg R_e^s$. In Eq. (4.40b), the term within the brackets becomes positive, thus resulting in a negative sensitivity value. In turn, a negative sensitivity means that if the design variable value is increased, i.e. the material composition is shifted toward concrete, the compliance will decrease. This complies with the aim of minimizing the objective function, thus an optimization algorithm would assign more concrete to the element in this example, just as it is desired in the case of compression-dominant stress states.

In fact, however, the Young's modulus of concrete is five to seven times lower than that of steel and its Poisson's ratio ($v_c = 0.2$) equals two-thirds ($v_s = 0.3$). The question, therefore, arises as to when a compressive stress state is "sufficiently" dominant so that an optimization algorithm prefers concrete over steel despite the former's lower stiffness. From a mathematical point of view, the threshold value R_e^* for which the sensitivity with respect to φ_e becomes zero is to be found. Utilizing $R_e^s = 1 - R_e^c$ from Eq. (4.45), then follows for a fixed χ_e:

$$\mathbf{u}_e^\top \left[R_e^* \mathbf{k}_e^c - (1 - R_e^*) \mathbf{k}_e^s \right] \mathbf{u}_e \overset{!}{=} 0 \tag{4.47a}$$

$$\Leftrightarrow R_e^* = \frac{\mathbf{u}_e^\top \mathbf{k}_e^c \mathbf{u}_e}{\mathbf{u}_e^\top \mathbf{k}_e^c \mathbf{u}_e + \mathbf{u}_e^\top \mathbf{k}_e^s \mathbf{u}_e} \tag{4.47b}$$

Thus, if $R_e^c > R_e^*$, then a stress state sufficiently dominant in compression is prevailing, causing the modified sensitivity to push the design variable update toward concrete. In the opposite case, namely $R_e^c < R_e^*$, the evolution of φ_e is driven toward steel.

To avoid numerical difficulties, it is necessary to apply numerical regularization prior to the design variable update, e.g. by filtering techniques. A suitable bi-material variant is presented in the following section.

4.3.3 Bi-material Filtering

The bi-material topology optimization approach also faces the well-known numerical challenges, namely checkerboards (Figure 4.10) and mesh dependency. A sensitivity filter adapted to the bi-material SIMP approach is given in [25]. The MRM-modified sensitivities are adjusted in dependence of the elemental sensitivities lying within the predefined filter radius r_{min}:

$$\left.\frac{\partial \tilde{f}(\chi, \varphi)}{\partial \chi_e}\right|_{mod} = \frac{1}{\chi_e \sum_{i=1}^{N_e} H_{ei}} \sum_{i=1}^{N_e} H_{ei} \chi_i \left.\frac{\partial f(\chi, \varphi)}{\partial \chi_i}\right|_{mod} \tag{4.48a}$$

$$\left.\frac{\partial \tilde{f}(\chi, \varphi)}{\partial \varphi_e}\right|_{mod} = \frac{1}{\sum_{i=1}^{N_e} H_{ei}} \sum_{i=1}^{N_e} H_{ei} \left.\frac{\partial f(\chi, \varphi)}{\partial \varphi_i}\right|_{mod} \tag{4.48b}$$

where

$$H_{ei} = \begin{cases} r_{min} - \Delta_{ei} & \text{if } \Delta_{ei} \leq r_{min} \\ 0 & \text{else} \end{cases} \tag{4.49}$$

and Δ_{ei} is the distance between the centroids of element e and i.

From a practical point of view, it may be reasonable to specify different filter radii (=minimum strut thicknesses) for concrete and steel due to different manufacturing constraints. For this purpose, an alternative filter can be defined, where the sensitivities with respect to the density variables are filtered depending on the element's material composition:

$$\left.\frac{\partial \tilde{f}(\chi, \varphi)}{\partial \chi_e}\right|_{mod} = \varphi_e \left[\frac{1}{\chi_e \sum_{i=1}^{N_e} H_{ei}(r_{min}^c)} \sum_{i=1}^{N_e} H_{ei}(r_{min}^c) \chi_i \left.\frac{\partial f(\chi, \varphi)}{\partial \chi_i}\right|_{mod} \right]$$

$$+ (1 - \varphi_e) \left[\frac{1}{\chi_e \sum_{i=1}^{N_e} H_{ei}(r_{min}^s)} \sum_{i=1}^{N_e} H_{ei}(r_{min}^s) \chi_i \left.\frac{\partial f(\chi, \varphi)}{\partial \chi_i}\right|_{mod} \right]. \tag{4.50}$$

Now, $H_{ei}(r_{min}^c)$ and $H_{ei}(r_{min}^s)$ are dependent on the respective filter radius for concrete and steel, respectively. If an element consists entirely of one material, the filter reduces to the special case given in Eq. (4.48a).

Figure 4.10 Bi-material topology optimization: (a) checkerboards, (b) result with filter.

4.3.4 Solving

Apart from the common nonlinear optimization algorithms, an efficient update scheme based on OC can likewise be developed for the bi-material approach [6]. For this purpose, first, the Lagrangian function of the optimization problem of Eq. (4.37) must be set up. It reads

$$L = \mathbf{U}^\mathsf{T}\mathbf{K}(\chi, \varphi)\mathbf{U} + \Lambda \left(V - \beta V^0\right) + \sum_{e=1}^{N_e} \lambda_e \left(10^{-3} - \chi_e\right)$$
$$+ \sum_{e=1}^{N_e} \gamma_e \left(\chi_e - 1\right) + \sum_{e=1}^{N_e} \delta \left(0 - \varphi_e\right) + \sum_{e=1}^{N_e} \zeta_e \left(\varphi_e - 1\right)$$

(4.51)

where $\Lambda \geq 0$, $\lambda_e \geq 0$, $\gamma_e \geq 0$, $\delta_e \geq 0$, and $\zeta_e \geq 0$ are the Lagrange multipliers for $e \in [1, N_e]$. Two stationarity conditions can now be defined with respect to the design variables. The first reads

$$\frac{\partial L}{\partial \chi_e} = \left(\frac{\partial \mathbf{U}^\mathsf{T}}{\partial \chi_e}\mathbf{K}\mathbf{U} + \mathbf{U}^\mathsf{T}\frac{\partial \mathbf{K}}{\partial \chi_e}\mathbf{U} + \mathbf{U}^\mathsf{T}\mathbf{K}\frac{\partial \mathbf{U}}{\partial \chi_e}\right) + \Lambda v_e - \lambda_e + \gamma_e = 0 \quad (4.52)$$

whereby the second becomes

$$\frac{\partial L}{\partial \varphi_e} = \left(\frac{\partial \mathbf{U}^\mathsf{T}}{\partial \varphi_e}\mathbf{K}\mathbf{U} + \mathbf{U}^\mathsf{T}\frac{\partial \mathbf{K}}{\partial \varphi_e}\mathbf{U} + \mathbf{U}^\mathsf{T}\mathbf{K}\frac{\partial \mathbf{U}}{\partial \varphi_e}\right) - \delta_e + \zeta_e = 0 \quad (4.53)$$

Similar to the procedure for the original SIMP described in Section 4.1.4, two optimality conditions can be derived, one for each design variable. However, the modified (and filtered) sensitivities must be considered here in order to take into account the compression and tension affinity of concrete and steel, respectively. The OC for the density variables does not differ from the original SIMP approach and takes the form:

$$\underbrace{\frac{-\frac{\partial f(\chi,\varphi)}{\partial \chi_e}\Big|_{\text{mod}}}{2\Lambda v_e}}_{X_e} = 1 - \frac{\lambda_e}{\Lambda v_e} + \frac{\gamma_e}{\Lambda v_e} \quad (4.54)$$

Consequently, the update scheme is also equal to the original approach and reads

$$\chi_e^{(k+1)} = \begin{cases} \left[X_e^{(k)}\right]^{0.5}\chi_e^{(k)} & \text{if } M_\chi^- \leq \left[X_e^{(k)}\right]^{0.5}\chi_e^{(k)} \leq M_\chi^+ \\ M_\chi^- & \text{if } M_\chi^- \geq \left[X_e^{(k)}\right]^{0.5}\chi_e^{(k)} \leq M_\chi^+ \\ M_\chi^+ & \text{if } M_\chi^- \leq \left[X_e^{(k)}\right]^{0.5}\chi_e^{(k)} \geq M_\chi^+ \end{cases} \quad (4.55)$$

where

$$M_\chi^- = \max\left\{(1-\mu)\,\chi_e^{(k)}, 10^{-3}\right\} \quad (4.56a)$$

$$M_\chi^+ = \min\left\{1, (1+\mu)\,\chi_e^{(k)}\right\} \quad (4.56b)$$

with $\mu = 0.2$ being the move limit.

By rearranging Eq. (4.53), an OC can also be specified for the second design variable:

$$\underbrace{-\left.\frac{\partial f(\chi,\varphi)}{\partial \varphi_e}\right|_{\text{mod}}}_{\Phi_e} = 2\left(\zeta_e - \delta_e\right) \tag{4.57}$$

Similarly, by analyzing the above equation concerning variations of φ_e, an update scheme can be derived also in this case. For further details, reference is made to [6], where it is shown that Φ_e has to approach zero in order to satisfy the optimality condition. Depending on whether R_e^c is greater or lower than the threshold value R_e^* from Eq. (4.47b), these two cases have to be distinguished in order to adequately take into account the element's prevailing stress state.

In the case $R_e^c > R_e^*$ ($\Phi_e > 0$), the following update scheme leads to an adequate evolution of the design variables:

$$\varphi_e^{(k+1)} = \begin{cases} \left[\alpha\Phi_e^{(k)} + 1\right]\varphi_e^{(k)} & \text{if } M_\varphi^- \leq \left[\alpha\Phi_e^{(k)} + 1\right]\varphi_e^{(k)} \leq M_\varphi^+ \\ M_\varphi^- & \text{if } M_\varphi^- \geq \left[\alpha\Phi_e^{(k)} + 1\right]\varphi_e^{(k)} \leq M_\varphi^+ \\ M_\varphi^+ & \text{if } M_\varphi^- \leq \left[\alpha\Phi_e^{(k)} + 1\right]\varphi_e^{(k)} \geq M_\varphi^+ \end{cases} \tag{4.58}$$

where

$$M_\varphi^- = \max\left\{(1-\mu)\,\varphi_e^{(k)}, 0\right\} \tag{4.59a}$$

$$M_\varphi^+ = \min\left\{1, (1+\mu)\,\varphi_e^{(k)}\right\} \tag{4.59b}$$

Here, $\alpha = 100$ is an "acceleration" factor, which counteracts a severe deceleration of convergence the closer Φ_e approaches zero. In addition, 1 is added to the term within the brackets for the case where $0 \leq \alpha\Phi_e^{(k)} < 1$ holds true.

In the opposite case, for $R_e^c < R_e^*$ ($\Phi_e < 0$), the following applies instead:

$$\varphi_e^{(k+1)} = \begin{cases} \left[\frac{1}{\left|\alpha\Phi_e^{(k)}-1\right|}\right]\varphi_e^{(k)} & \text{if } M_\varphi^- \leq \left[\frac{1}{\left|\alpha\Phi_e^{(k)}-1\right|}\right]\varphi_e^{(k)} \leq M_\varphi^+ \\ M_\varphi^- & \text{if } M_\varphi^- \geq \left[\frac{1}{\left|\alpha\Phi_e^{(k)}-1\right|}\right]\varphi_e^{(k)} \leq M_\varphi^+ \\ M_\varphi^+ & \text{if } M_\varphi^- \leq \left[\frac{1}{\left|\alpha\Phi_e^{(k)}-1\right|}\right]\varphi_e^{(k)} \geq M_\varphi^+ \end{cases} \tag{4.60}$$

For detailed information, see [6], where numerical studies prove the superiority of the update schemes over utilizing nonlinear optimization algorithms with default parameters in terms of robustness, convergence speed, and numerical reliability.

4.3.5 Optimization Process

Figure 4.11 shows the flow chart of the bi-material topology optimization approach. After initializing the model, a finite element calculation with elastic material properties is performed. From the resulting displacement field, the principal stresses and strains are then determined. They are utilized to compute the modification factors

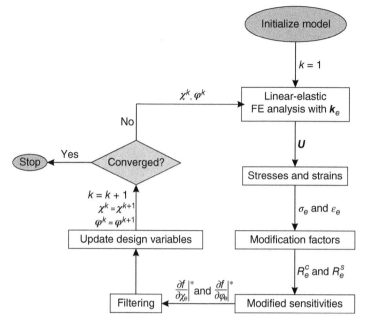

Figure 4.11 Flow chart of the bi-material topology optimization approach.

of each element. Finally, the modification factors are applied to modify the sensitivities of the objective function. They are then filtered and finally employed by the optimizer, e.g. the OC-based update scheme, to evolve the design variables. At the end of the iteration, a check is made whether the convergence criterion is fulfilled or not. If yes, the optimization process ends. If not, a new optimization loop is initiated by performing a linear elastic FE analysis. A possible stop criterion is the largest difference between the design variables of two subsequent iterations:

$$\max \Delta x_e = \max \left| x_e^{(k+1)} - x_e^{(k)} \right| \leq \mathrm{tol}_x \tag{4.61}$$

where $\mathrm{tol}_x = 10^{-2}$ is a reasonable limit [31].

4.4 Examples

4.4.1 One-material Structures

Example 4.1 (Variation of volume fraction). The influence of the volume fraction on the optimization result in one-material topology optimization is demonstrated in Figure 4.12. Here, the examined subject is a single span beam, loaded by a single load $F = 200$ kN. Length and height values are given in [m]. The design space consists of $240 \times 40 = 9600$ quadrilateral elements. The material parameters are given by $E = 28\,000$ MPa for the Young's modulus and $v = 0.20$ for the Poisson's ratio. The penalty exponent takes the typical value $p = 3$ and a sensitivity filter with $r_{\min} = 1.5\,a_e$ is applied for numerical regularization, where a_e represents an

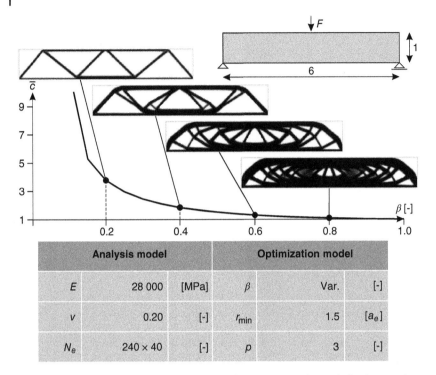

Figure 4.12 Influence of the employed material amount on the optimization results.

Analysis model			Optimization model		
E	28 000	[MPa]	β	Var.	[-]
v	0.20	[-]	r_{min}	1.5	$[a_e]$
N_e	240 × 40	[-]	p	3	[-]

element length. The share β of available material compared to the initial volume is varied.

As the amount of material increases, the resulting structure also changes qualitatively by developing additional struts. At the same time, however, a nonlinear relationship between structural compliance and material usage becomes apparent. Figure 4.12 demonstrates this by means of the normalized compliance values for different magnitudes of β, which are related to the compliance for a completely filled design space ($\beta = 1.0$). While at very low values for β, little variation in the material input has great influence on the structural compliance, the influence of the material amount on c decreases significantly with increasing volume fractions. In other words, this means that a material reduction of up to 40 % hardly affects the structure's stiffness in this example. However, reductions of up to about 20 % also appear reasonable, since here the material savings still outweigh the loss of stiffness.

Example 4.2 (Variation of the filter radius). The influence of the filter radius (sensitivity filter) on the results in one-material topology optimization using the example of a single-span beam loaded with $F = 200$ kN from the previous section is shown in Figure 4.13. Both the discretization of the design space with 240 × 40 = 9600 quadrilateral elements as well as the employed material parameters, $E = 28\ 000$ MPa and $v = 0.20$, remain unchanged from the previous example. The penalty exponent equals $p = 3$ and the material amount is limited by $\beta = 0.60$.

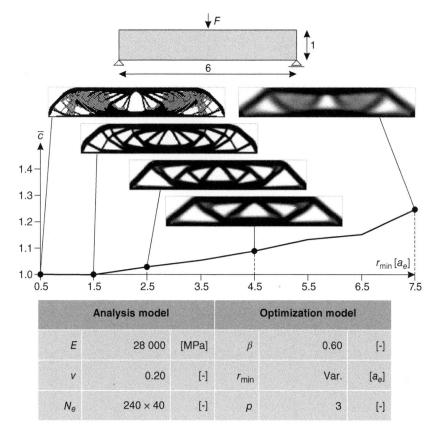

Analysis model			Optimization model		
E	28 000	[MPa]	β	0.60	[-]
v	0.20	[-]	r_{min}	Var.	$[a_e]$
N_e	240 × 40	[-]	p	3	[-]

Figure 4.13 Influence of the sensitivity filter radius on one-material topology optimization results.

The diagram shows the change of the compliance value as a function of the filter radius, which is defined in relation to an element length a_e. The compliance is scaled to that of the unfiltered result ($r_{min} = 0.5a_e$) and increases continuously for larger filter radii. Furthermore, the checkerboard effect becomes apparent in the unfiltered result. There are two main reasons for the increase in compliance. On the one hand, the initially thin struts are merged to form increasingly thicker ones, which has a negative effect on the stiffness of the structure. On the other hand, large filter radii lead to a growing blurring effect, i.e. the gradual decrease of the design variable values over larger areas around a strut. These gray structural parts of intermediate relative densities lower the overall structure's stiffness. However, it also becomes obvious that filtering by means of moderate radii compared to the mesh size, both the complexity of the results and the minimum strut thickness can be controlled without excessively compromising the stiffness.

Example 4.3 (Variation of material parameters). In this example, the influence of the Young's modulus on the one-material topology optimization results will be demonstrated. The examined structure is the simple beam with point load from the previous examples. The parameters used are $F = 200$ kN, $N_e = 240 \times 40 = 9600$, $v = 0.20$, $r_{min} = 1.5a_e$, $p = 3$, and $\beta = 0.40$.

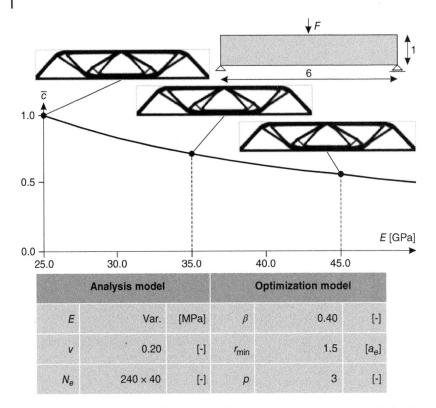

Analysis model			Optimization model		
E	Var.	[MPa]	β	0.40	[-]
v	0.20	[-]	r_{min}	1.5	$[a_e]$
N_e	240 × 40	[-]	p	3	[-]

Figure 4.14 Influence of the Young's modulus on one-material topology optimization results.

Figure 4.14 shows the results of the parametric study. The compliance is plotted over the Young's modulus. Apparently, there is a slightly nonlinear relationship between material and structural stiffness. In contrast, with otherwise identical boundary conditions, the optimized material distribution is independent of the Young's modulus. Hence, E only affects the quantitative evaluation of the optimization, namely when the exact compliance value is sought, yet not the fundamental optimization result.

Example 4.4 (Form finding of bridge pylons 1). The pylon geometry of a cable-stayed bridge is sought. The design space is specified according to Figure 4.15a with $h/b = 100/40$ [m] and a block-out of $\Delta h/\Delta_b = 6/24$ [m] at $h' = 25$ m. The two support points are defined in the quarter points of the lower border of the design space. The space required for the roadway is defined as non-design domain and excluded from the optimization. The cable forces are aggregated into a single vertical force that acts on the center of the design space's upper boundary. Since the single force is the only, a dimensionless unit load is used for simplification, which makes no difference to the resulting material distribution. The FE mesh

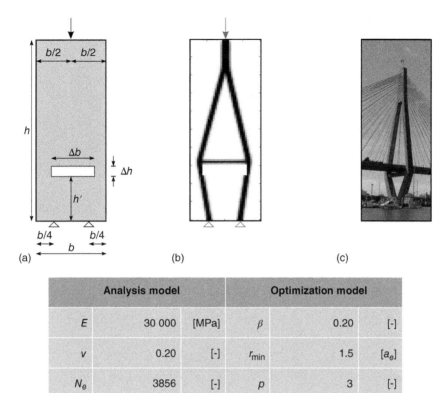

	Analysis model			Optimization model	
E	30 000	[MPa]	β	0.20	[-]
v	0.20	[-]	r_{min}	1.5	$[a_e]$
N_e	3856	[-]	p	3	[-]

Figure 4.15 Pylon 1 example: (a) design space, (b) optimization result, (c) pylon of the Anzac bridge in Sydney, Australia [32].

consists of $40 \times 100 - 24 \times 6 = 3856$ elements. As a volume constraint, $\beta = 0.20$ is applied and the sensitivity filter radius is set to $r_{min} = 1.5\, a_e$.

The optimization result is shown in Figure 4.15b. At the upper section, the pylon takes the load via a vertical component, then splits into two separate struts, which spread to the outer dimensions of the non-design space at the roadway. Above the clearance gauge, the two pylon sections are braced via a lateral strut, which is connected to the lower supports by two inclined pillars. The material distribution resembles, for instance, the RC pylons of the Anzac cable-stayed bridge over Johnstons Bay in Sydney, Australia, depicted in Figure 4.15c. In this case, however, the lateral strut is underneath the roadway, since it also serves as a bearing for it.

Example 4.5 (Form finding of bridge pylons 2). The influence of a modified design space on the pylon geometry is investigated in this example. The support conditions as well as the clearance gauge are unchanged from the Pylon 1 example. However, the cable loads are idealized here by two vertical forces on the upper side of the design space, which act at the quarter points (Figure 4.16a). Both loads are assumed equal. FE mesh, volume constraint, and sensitivity filter radius are identical to the prior example.

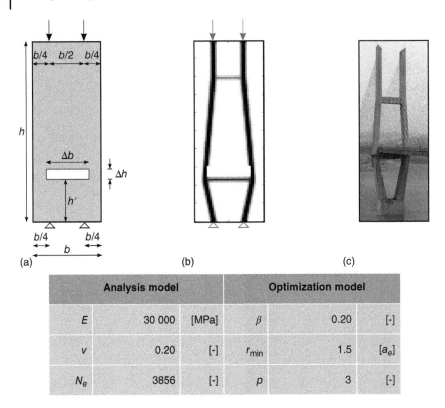

	Analysis model			Optimization model	
E	30 000	[MPa]	β	0.20	[-]
v	0.20	[-]	r_{min}	1.5	$[a_e]$
N_e	3856	[-]	p	3	[-]

Figure 4.16 Pylon 2 example: (a) design space, (b) optimization result, (c) Lidu bridge in China [32].

The resulting pylon structure in Figure 4.16b now consists entirely of two vertical substructures on the left and right, which are joined together by transverse braces wherever the struts are deflected: one lateral strut is located approximately in the upper fifth of the design space and the other below the non-design space of the roadway slab. The optimization result resembles the pylons of, for example, the Lidu road bridge over the Yangtze river in China depicted in Figure 4.16c.

Example 4.6 (Conceptual bridge design 1). The design space of a bridge crossing a river is shown in Figure 4.17a with $h/b = 40/160$ [m]. On the lower side, a non-design domain with $\Delta h/\Delta b = 20/70$ [m] is defined to provide a passage for shipping traffic. The support points are placed at sufficient distance from the river banks. The roadway slab is defined as non-design space by a predetermined row of elements with full material allocation. A uniformly distributed load acts on the upper side of the deck. The design space is discretized by $160 \times 40 - 70 \times 20 = 5000$ quadrilateral elements. The volume fraction is set to $\beta = 0.20$, while $r_{min} = 1.5\ a_e$ is applied as sensitivity filter radius.

The topology optimization result is shown in Figure 4.17b. An arch structure forms between the bearings and intersects the roadway slab. Vertical tension members

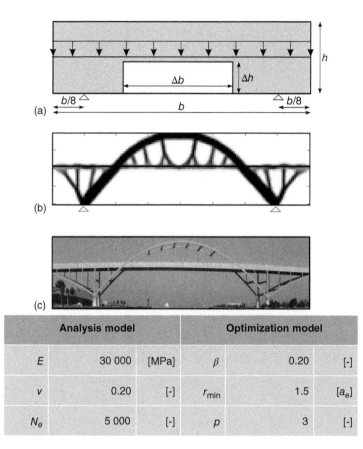

	Analysis model				Optimization model	
E	30 000	[MPa]	β		0.20	[-]
v	0.20	[-]	r_{min}		1.5	$[a_e]$
N_e	5 000	[-]	p		3	[-]

Figure 4.17 (a) Design space, (b) optimization result, (c) Daniel Hoan Memorial Bridge in the United States [32].

connect the arch with the latter. At the bridge ends, in contrast, the deck is supported by compression components linked to the supports. The design proposal bears sim-ilarities to, for example, the Daniel Hoan Memorial steel tied arch bridge, which spans the Milwaukee River in the United States (Figure 4.17c). It is characterized by minimizing bending effects and encouraging pure axial stresses.

Example 4.7 (Conceptual bridge design 2). A bridge design space similar to the previous example is shown in Figure 4.18a. The aim this time is to span a valley, hence the position of the fixed supports is altered ($h' = 8$ m). Again, the roadway slab as well as a non-design space with $\Delta h/\Delta b = 20/86$ [m] are predefined and thus excluded from the optimization. The size of the design space is $h/b = 40/120$ [m], the volume is constrained to $\beta = 0.20$, and the sensitivity filter radius equals $r_{min} = 1.5\ a_e$.

Figure 4.18b depicts the resulting material distribution that can be interpreted as a through arch bridge with hangers. A built example resembling this design proposal is

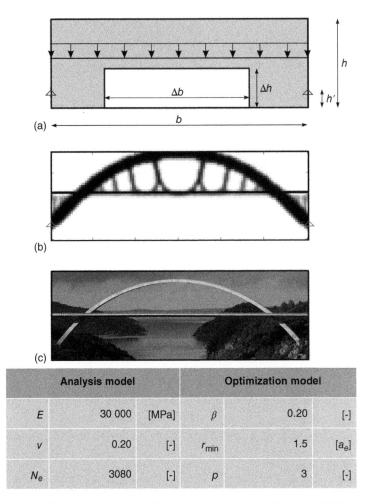

(a)

(b)

(c)

Analysis model			Optimization model		
E	30 000	[MPa]	β	0.20	[-]
ν	0.20	[-]	r_{min}	1.5	$[a_e]$
N_e	3080	[-]	p	3	[-]

Figure 4.18 (a) Design space, (b) optimization result, (c) Svinesund Bridge connecting Sweden and Norway [32].

the Svinesund Bridge which crosses the Iddefjord, thus connecting Sweden and Norway (Figure 4.18c) [33].

Example 4.8 (Multi-span girder). A bridge design with three fields, each of equal size, loaded by a uniformly distributed load is sought, see Figure 4.19a. The roadway slab is given as non-design domain, which is represented by one row of elements that are assigned full material. The supports are placed right below the slab. Design space is provided above as well as underneath the roadway.

Figure 4.19b shows the optimization result for $\beta = 0.20$ and $r_{min} = 1.5$. The outer fields are each spanned by an arch, where the roadway is connected by hangers. At the inner bearings, the arches move below the deck and thus remain compressed, however, they exhibit a lower rise than on the outer fields. At midspan, the two

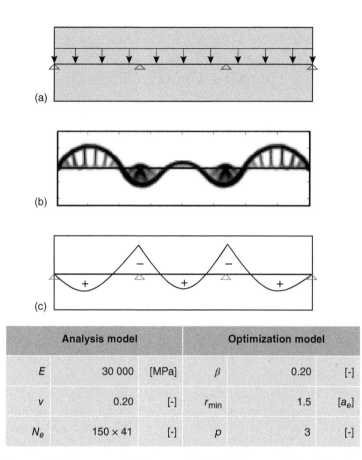

Figure 4.19 Three-span bridge example: (a) design space, (b) optimization result, (c) bending moment [32].

	Analysis model			Optimization model	
E	30 000	[MPa]	β	0.20	[-]
v	0.20	[-]	r_{min}	1.5	$[a_e]$
N_e	150 × 41	[-]	p	3	[-]

mirrored arch substructures join to form an arch pointing upward again, which is predominantly located below the roadway slab. The wave-like structure obviously forms affine to the bending moment and counteracts it as becoming evident by Figure 4.19c. So the bending effect is proportionally resolved into axially stressed members of purely tensile and compressive stresses.

Example 4.9 (Multiple load cases). In this example, the influence on the optimization results when considering different load cases is demonstrated using a simply supported wall with a width to height ratio of 3/1. The wall is loaded by point loads F_1, F_2, and F_3 at each quarter point, respectively. All loads F_i are of the same extent. The design space is composed of 192 × 64 quadrilateral elements, each with an element length of about $a_e = 0.015$. For the optimization, $\beta = 0.30$ and sensitivity filtering with $r_{min} = 1.5$ a_e is applied.

Figure 4.20a shows the results when F_1, F_2, and F_3 act separately, while Figure 4.20b depicts the material distribution when all three forces are applied

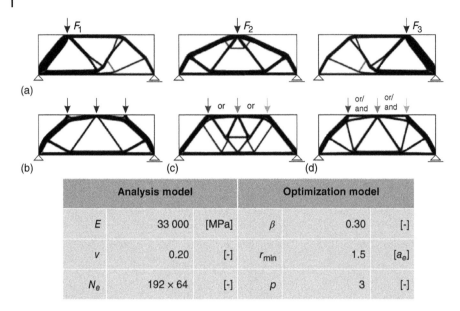

Figure 4.20 Wall with three point loads: (a) acting individually, (b) acting together, (c) as exclusive load cases, (d) as inclusive load cases.

Analysis model			Optimization model		
E	33 000	[MPa]	β	0.30	[-]
v	0.20	[-]	r_{min}	1.5	$[a_e]$
N_e	192 × 64	[-]	p	3	[-]

simultaneously. However, if the loads are independent of each other, the latter is not the optimal result, as we can assume from the different results in (a). To consider the loads in different load cases, Figure 4.20c gives the outcome resulting from the approach in Section 4.1.6. Here it is assumed that the load cases are exclusive, i.e. either F_1 or F_2 or F_3 (three load cases) can act. The result differs from that given in Figure 4.20b because the structure is obviously more robust against the varying load applications points. Figure 4.20d shows the resulting material distribution for the case of inclusively defined forces, which means that F_1, F_2, and F_3 can act either individually as in Figure 4.20c or together in arbitrary combinations. This yields a total of seven load cases, namely F_1, F_2, F_3, $F_1 + F_2$, $F_1 + F_3$, $F_2 + F_3$, and $F_1 + F_2 + F_3$. Again, the result differs, but similar to Figure 4.20b, with additional bracing struts being included to link the outer load application points with the structure. Hence, it is evident that the accurate definition of load cases is crucial for the optimization outcome to ensure robust designs.

Example 4.10 (Two load cases). The design space of a cantilever arm loaded by a vertical and a horizontal load F and H, respectively, is depicted in Figure 4.21a. The length and height of the design space have a ratio of 3 to 1. For optimization, 180 × 60 quadrilateral elements of the size $a_e \times a_e$ are used. The volume fraction is limited to $\beta = 0.2$ and a sensitivity filter with $r_{min} = 1.5\,a_e$ is employed for regularization.

Figure 4.21b shows the optimization results in case of F or H acting individually, while Figure 4.21c represents the material distributions when the loads are applied together for three different ratios of F/H. The latter determines the

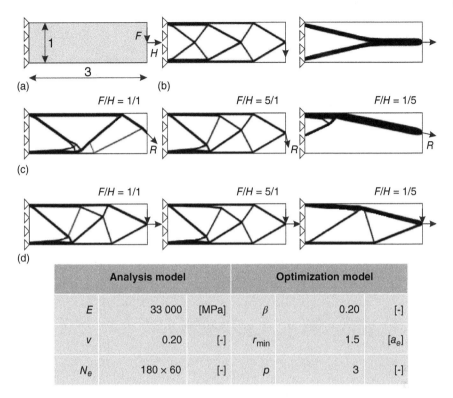

Figure 4.21 (a) Design space of a cantilever with vertical and horizontal loads and results for (b) separate loading, (c) simultaneous loading, and (d) inclusive load cases.

inclination of the resulting force $R = \sqrt{F^2 + H^2}$, which in turn affects the structure. Very steep and very flat resultants lead to outcomes which resemble the material distributions resulting from loading containing only F and H, respectively. In contrast, Figure 4.21d shows the results when defining the loads as different including load cases. Three load combinations are possible, namely F, H, and $F + H$. Again it is evident that the definition of load cases apparently has significant impact on the material distribution.

4.4.2 One-material Stress-biased Structures

Example 4.11 (**Material steering**). The design problem of a three-span bridge with varying span widths is shown in Figure 4.22. A uniformly distributed load $q = 100$ kN/m acts on the roadway and two piers as well as the supports are predefined. The roadway slab and the piers consist of solid elements with full material allocation, which cannot be altered by the optimization algorithm. The design space of 180 times 80 meters is discretized by 180×80 quadrilateral elements of the size $a_e = 1$ m and a filter with radius $r_{min} = 1.5\, a_e$ is utilized for numerical regularization. The permitted overall volume is constrained by $\beta = 0.1$.

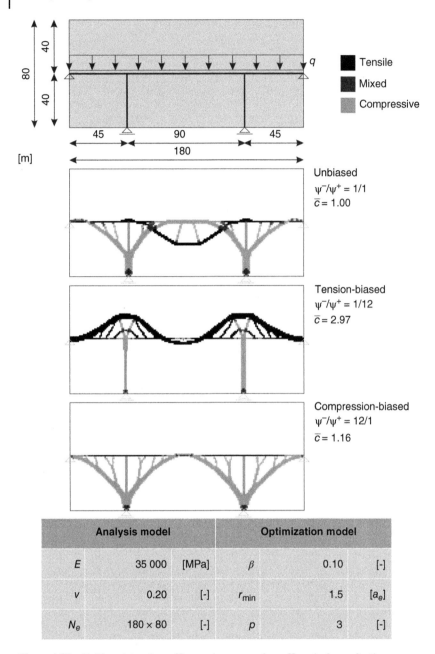

Figure 4.22 Unbiased, tension-affine and compression-affine designs of a three-span bridge, according to [22].

Figure 4.22 presents three optimized material distributions: one for the unbiased approach and two additional ones for compression affinity ($\Psi^-/\Psi^+ = 12/1$) and tension affinity ($\Psi^-/\Psi^+ = 1/12$), respectively. The results are given together with their corresponding compliance values scaled to that of the unbiased result. The unbiased result shows a combination of compressive arches and tensile substructures for the roadway at midspan. The overall design consists of 78 % compressive, 18 % tensile, and 4 % mixed elements, whereby the latter are elements exhibiting stress states within the range $-1 \le s_e \le -0.5$. Hence, the resulting optimized design is naturally compression-oriented, with the tensile elements providing additional stiffness.

The tension-biased design increases the proportion of tensile elements to 54 %, whereas the compressive elements drop to only 34 % and a 12 % proportion of mixed elements remains. The structure now resembles a suspension bridge. However, the approximately three times as high compliance value shows that it differs greatly from the natural load-bearing behavior of the unbiased approach. In contrast, in the compression-biased case, the proportion of compressive elements increases to 91 %. The elements being predominantly under tension (2 %) as well as the mixed elements (7 %) vanish almost from the design. The structure transforms into a pure compressive arch, where merely parts of the roadway slab exhibit tension imposed by bending. The compliance value increases moderately by 16 % compared to the unbiased design, since the additional stiffening tensile substructure at midspan is no longer increasing the bridge's stiffness.

4.4.3 Bi-material Structures

Example 4.12 (Material variation in bi-material design). The effect of different Young's modulus ratios in bi-material topology optimization is illustrated in Figure 4.23. A simple beam with a concentrated load at midspan and a length to height ratio of 6/1 is taken as an example. The design space is discretized with 240×40 quadrilateral elements. The Young's moduli should have ratios of 1/1, 4/1, and 1/4, where the ratio 1/4 corresponds to the expected ratio of high-strength concrete E_c ($\sim 50\,000$ MPa) relative to conventional steel E_s ($\sim 200\,000$ MPa). Concrete is treated as compression-only material with $v_c = 0.2$ and steel as tension-only material with $v_s = 0.3$. For the optimization, a sensitivity filter adopting $r_{min} = 1.5\,a_e$ is applied, where a_e is the element length. For the volume constraint, $\beta = 0.30$ is set as a limit.

The optimization results show that with balanced Young's moduli, the resulting structure has a similar ratio of material amounts, with slightly more concrete being used, which may be due to its lower Poisson's ratio. At a greater Young's modulus for steel, a truss structure forms with significantly thicker steel struts than concrete elements, whereas the opposite applies in the case where $E_c > E_s$. Thus, in this example, the Young's moduli have a clear influence on the optimization results since the stiffer material is favored due to its higher contribution to the overall stiffness of the structure.

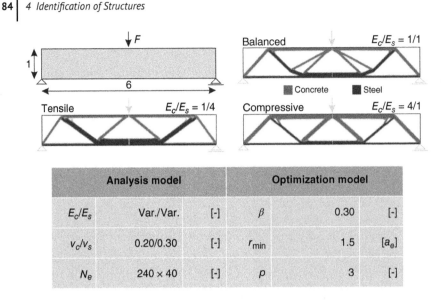

	Analysis model				Optimization model	
E_c/E_s	Var./Var.	[-]	β		0.30	[-]
v_c/v_s	0.20/0.30	[-]	r_{min}		1.5	$[a_e]$
N_e	240×40	[-]	p		3	[-]

Figure 4.23 Bi-material optimized structures for different ratios of the Young's moduli.

Example 4.13   (Filter radius with bi-material design).   The design space of a single span beam with a concentrated load F acting on its lower edge is shown in Figure 4.24. The design space dimensions are $L/H = 2/1$ and the Young's moduli ratio equals $E_c/E_s = 1/4$, which might apply to the relation of ultra-high performance concrete to structural steel. The Poisson's ratios are $v_c = 0.2$ and $v_s = 0.3$, respectively. 160×80 quadrilateral elements discretize the design space and $\beta = 0.3$ is applied as the volume constraint limit. The influence of different sensitivity filter radii for the materials (see Section 4.3.3) is to be investigated. In doing so, no filtering of the sensitivities with respect to the phase variables φ_e is performed.

The results in Figure 4.24 illustrate that with applying different filter radii the minimum strut thicknesses can be controlled for each of the two materials. Using an identical filter radius yields a concrete arch with a total of eight steel "spokes" connecting it with the acting force at the lower edge. The arch has a much more pronounced thickness than these tension elements ("spokes"), which is due to the fact that the employed concrete has a Young's modulus that equals 1/4 that of steel. Consequently, about four times as much material is applied to activate comparable stiffness. By increasing the filter radius for steel (r_{min}^s) by ~ 3.5 times, the optimized structure exhibits fewer, but thicker steel elements, whose size is comparable to those made of concrete. Due to the reduction in the number of spokes, the result transforms from an arch to a truss structure. In contrast, the opposite case, namely increasing the filter radius for concrete (r_{min}^c), has hardly any influence on the result in this particular example, since the difference in the material stiffnesses roughly corresponds to r_{min}^c. This demonstrates that although the filter radii can be used to limit the *lower* limit of the structural element thicknesses, no influence can be exerted on the actual strut dimensions.

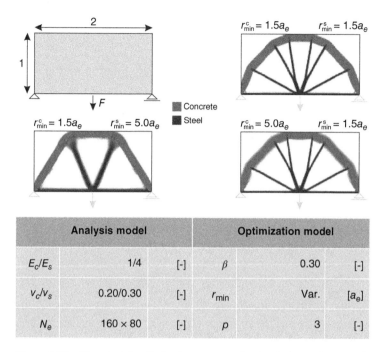

Figure 4.24 Bi-material optimized structures of a simple beam for material-dependent sensitivity filter radii.

Example 4.14 (Bi-material multi-span girder). A composite bridge structure across a river is sought. The initial planning situation along with the terrain showing its respective constraint points is depicted in Figure 4.25. The terrain allows a three-span bridge, whereby its structure can only be designed above the roadway slab, the position of which is predefined. Based on these considerations, the design space is modeled as shown in Figure 4.25b, with four supports and a given carriageway slab at the lower edge, considering of one row of fixed solid elements, which is exposed to a uniformly distributed load. It consists of 20 m height and 100 m length and it is subdivided into short outer spans (20 m) and a pronounced midspan (60 m) to cross the river. The FE mesh consists of 200×40 elements in total. The material parameters utilized are $E_c = 40\,000$ MPa and $v_c = 0.2$ for concrete and $E_s = 200\,000$ MPa and $v_s = 0.3$ for steel. For optimization, $\beta = 0.20$ is set and a sensitivity filter with $r_{min} = 1.5\,a_e$ is applied, where $a_e = 0.5$ m represents the size of an element edge.

The bi-material topology optimization result is depicted in Figure 4.25c. The inner field is spanned by a high compression arch made of concrete. The roadway is suspended by steel traction cables. Above the intermediate supports, concrete compressive struts form, which resemble pylons and transfer the loads from the fields via steel elements under tension to the bearings. Depending on the location, the roadway slab is either mainly under compressive or tensile stress. At midspan, tensile stresses dominate, hence it consists of steel here, whereas at the outer fields it is mainly subjected to compressive stresses and hence comprises concrete.

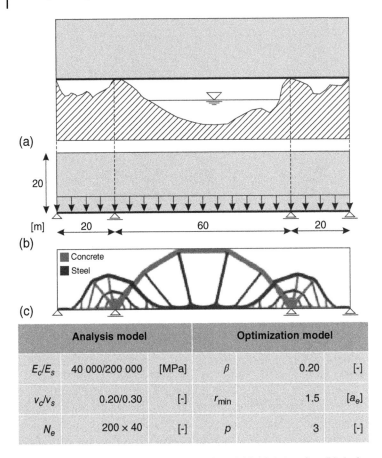

	Analysis model			Optimization model	
E_c/E_s	40 000/200 000	[MPa]	β	0.20	[-]
v_c/v_s	0.20/0.30	[-]	r_{min}	1.5	$[a_e]$
N_e	200 × 40	[-]	p	3	[-]

Figure 4.25 Composite river bridge problem: (a) initial situation, (b) design space, (c) bi-material topology optimization result.

Example 4.15 (Bi-material girder with stepped support). The optimized structure of a girder in a factory building between pillars at a distance of 24 m is sought. The supports are stepped upward to the center line of the design space. Figure 4.26a illustrates the initial situation. The loads are applied pointwise via purlins at a spacing of 2 m. For modeling the design space (Figure 4.26b), symmetry is exploited to reduce the computational effort. The height within which the structure can develop is set to 1 m each above and below the supports. In doing so, it allows the shape to adapt the bending moment in the best possible way. For the FE mesh, 180 × 30 quadrilateral elements of the size $a_e \times a_e$ with $a_e = 0.067$ m are employed. The materials adopted are concrete with $E_c = 36\,000$ MPa and $v_c = 0.2$, on the one hand, and steel with $E_s = 200\,000$ MPa and $v_s = 0.3$, on the other hand, as compression-only and tension-only material, respectively. For optimization, the volume constraint is set to $\beta = 0.35$ and the sensitivity filter radius equals $r_{min} = 1.5\,a_e$ for both materials.

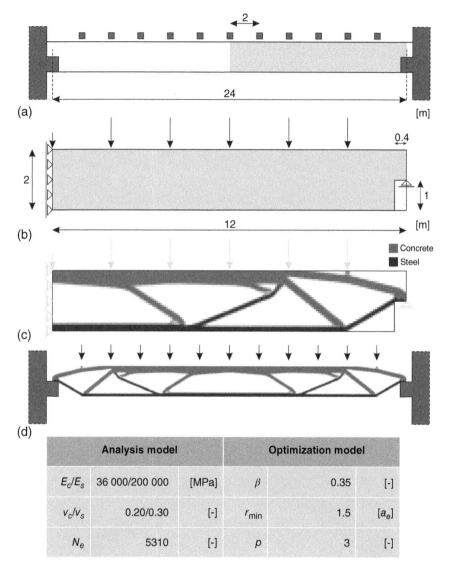

Figure 4.26 (a) Composite girder problem, (b) design space, (c) bi-material topology optimization result, (d) optimized girder of concrete (top) and steel (bottom).

The result is displayed in Figure 4.26c, d. The materials are arranged in a truss-like manner, with the compression struts consisting of concrete and the tension struts being made of steel. The trapezoidal area at midspan can be designed in such a way to provide bending stiffness for robustness against minor load scattering. The design space height is completely utilized. The structure's shape follows the bending moment with the compression and tension flange thicknesses decreasing from midspan to the supports. Contrary, the diagonal struts adapt to the shear force and increase in size.

Example 4.16 (Bi-material arch bridge). The topological optimization of a bridge structure over a valley, where the region above the roadway slab is to remain free, is shown in Figure 4.27. On the upper side of the design space, the roadway is therefore predefined as a solid element strip that must not be changed by the algorithm. The valley should also remain untouched to the most extent possible, therefore a void clearance gauge of 80 m × 15 m is specified. Fixed bearings are placed in the lower corners of the design space at a distance of 160 m from each other. The overall design space of 160 m × 50 m consists of $320 \times 100 - 160 \times 30 = 27\,200$ elements. The material parameters adopted are $E_c = 42\,000$ MPa and $v_c = 0.2$ for concrete and $E_s = 200\,000$ MPa and $v_s = 0.3$ for steel. Sensitivity filtering with $r_{min} = 1.5$ times an element length $a_e = 0.5$ m is applied for both materials. The residual volume ratio is aimed at $\beta = 0.25$.

The outcome of the topological optimization resembles a concrete arch. The structure is almost completely loaded with compressive stresses where only very few elements exhibit tension. This makes it ideal for a (reinforced) concrete design.

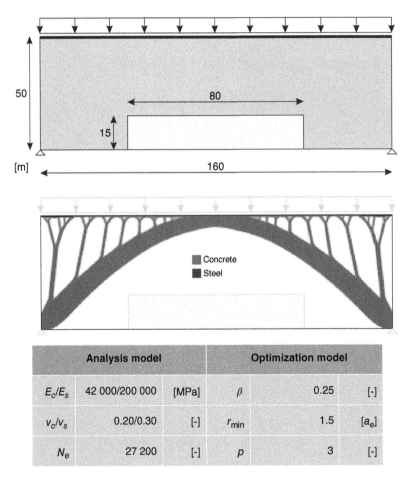

	Analysis model			Optimization model	
E_c/E_s	42 000/200 000	[MPa]	β	0.25	[-]
v_c/v_s	0.20/0.30	[-]	r_{min}	1.5	$[a_e]$
N_e	27 200	[-]	p	3	[-]

Figure 4.27 Composite arch bridge problem.

4.5   Applications

4.5.1   Solar Thermal Collectors

Optimized designs are particularly useful when components have high repetition rates. Then, even small reductions in a single element sum up to reasonable overall savings that equalize the increased efforts of optimization. Simply speaking, savings in materials, production, and transport outbalance the enhanced engineering.

Solar thermal collectors in solar fields of concentrated solar power (CSP) plants [34, 35] are examples for such structures that repeat each other hundreds of times. The collectors focus the incident solar radiation and are arranged in a line-like or chess-pattern manner.

Collectors can be roughly divided into line-focus and point-focus systems (Figure 4.28). Most established line-focus systems are parabolic troughs (a) that – arranged to rows – focus the solar radiation onto a line, namely a longitudinal absorber tube [36–38]. In comparison, the most established point-focus systems are solar power towers (b). In the circular solar field, single collectors, so-called heliostats [39, 40], are arranged around a central tower. The heliostats deflect the solar radiation onto a point at the top of the tower where the receiver is placed.

For maximum solar concentration, the mirror surfaces of the collectors, and thus their supporting structures, must follow a strict geometric shape of a parabola (line) and a paraboloid (point), respectively. These geometries must be fulfilled by the production (initial state) and kept during the daily sun tracking (tracking state). The latter requires a high stiffness so that deformations remain small in all relevant orientations of the collector. Up to now, both parabolic troughs and heliostats are mainly built up as steel frameworks with point-wise supported mirrors [41]. Here, alternative concrete solutions are derived and designed in a material minimized way using optimization. They benefit from the free shapeability of concrete, its availability as well as its lower costs compared to steel.

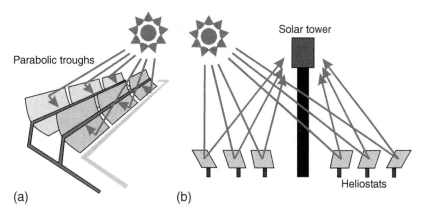

Figure 4.28   Principle of solar thermal collectors: (a) line-focusing parabolic troughs and (b) point-focusing solar tower with heliostats.

The design of the collectors aims at counteracting objectives, namely a maximum stiffness for solar concentration and a minimum use of materials motivated by resource efficiency. This means, a low-weight structure with minimal deformations, i.e. maximum stiffness, is sought. Here, a one-material CTO is used that targets on a minimum compliance (Section 4.1). At the same time, the volume is reduced proportional to the volume fraction β. In doing so, the two-objective problem is dissolved into the minimization of the compliance where the material reduction serves as a constraint.

Typical elements of the optimization process are

- A numerical simulation model – a standard finite element method (FEM) is used for this – of the structure that evaluates deformations and stresses in the step-wise form finding. Both values are restricted to acceptable limits. The deformation limit, strictly speaking a limitation of waviness, ensures a necessary exactness of solar ray concentration. The stress limit occurs due to use of concrete that should remain uncracked. To ensure this, concrete undergoes a certain tensile stress limit in the state of serviceability.
- Different loads cases of mainly dead loads and wind that arise from the different orientations of the collectors during the daily sun tracking ("morning position", so looking to the east, "noon position" to look upward and "evening position" when looking to west as well as intermediate ones like the often governing 45° positions).
- A superposition of relevant stress and deformation values that arise from the different load cases and must be checked at each location of the structure.

In the following applications, both parabolic troughs and heliostats are given.

It should be noted that other formulations of the optimization problem work too. One is to dissolve the optimization into the minimization of the volume whereby the low deformation, i.e. stiffness, demands serve as constraints. Hereby, shape optimization methods are the more efficient procedures [38, 42–45].

4.5.1.1 Parabolic Trough Collectors

The design focuses on concrete shells that act as the direct collectors supporting the mirroring surface. It is separated into two successive steps: first, a solid shell with a continuous inner and outer surface is derived. It is motivated from the idea of barreled roofs and merges a load-bearing structure and a continuous subsurface for the mirrors.

Second, the solid shell is deduced at its outer face to achieve further material savings. Doing so, stiffeners are introduced and the intermediate shell is diminished to a thin residual concrete layer. Optimization serves to derive the ideal pattern of stiffeners.

Materials and Actions A specific concrete is chosen with respect to two principal material properties. First, it needs a high Young's modulus to enhance axial and bending stiffness. Second, it must provide a high tensile strength to prohibit cracking that would provoke pronounced softening and thus an inacceptable loss of stiffness.

A high-performance concrete (HPC) based on the binder Nanodur compound 5941 [46] is chosen. It exhibits a Young's modulus of E_{cm} = 52 700 MPa that lies about 50 % beyond the one of NC and an axial tensile strength f_{ctm} of almost 10 MPa in the average. Relevant material properties of the HPC are summarized in Table 4.1.

The basic geometry of the shells is orientated at the most established parabolic trough collector module "EuroTrough" [41, 47, 48] with a length of 12 m and a width of 5.77 m. The focal length f, being the distance between the parabola vertex and the receiver, corresponds to 1.71 m and defines the curvature of the parabola. Figure 4.30 illustrates the geometry in ground view and cross section.

For the evaluations of stresses and deformations, dead loads and wind actions are taken into account. Both actions differ in the course of tracking.

Moreover, three basic situations of the wind speed have to be taken into account: first, an operational state, where the collector tracks the sun under low wind speeds of $v_{ref} \leq 10$ m/s. Second, a transition state, where the collector moves into a safety position under moderate wind conditions of 10 m/s $< v_{ref} \leq 15$ m/s, and third, a survival state, where the collector is hold in a stationary safety position under stormy or hurricane winds of up to $v_{ref} \leq 33$ m/s. In operational and transition state, the collector's orientation is variable. In survival state, the collector is fixed to the position that corresponds to the minimum wind pressure distribution being the so-called "noon position" (opening to the top). The wind loads w can be derived with respect to the gust pressure q_b by:

$$w = c_p \cdot q_b \tag{4.62}$$

with

$$q_b = \frac{1}{2} \rho_{air} v_{ref}^2 \tag{4.63}$$

Here, ρ_{air} denotes the air density of 1.25 kg/m³. The gust pressure exhibits q_b = 0.063 kN/m² for operation, 0.141 kN/m² for transition, and 0.681 kN/m² for survival state. Related wind pressure coefficients c_p are given, for example, in [38, 49–51].

Modules are usually assembled to a row of several single modules and then named collectors. One drive, either located at one end or intermediately, realizes sun tracking. Doing so, also torsional loads due to unbalanced or one-sided wind actions,

Table 4.1 Material properties of Nanodur concrete according to [42] (*m* – mean value)

Description		Value	Unit
Young's modulus	E_{cm}	52 700	MPa
Poisson's ratio	v	0.20	—
Compressive strength	f_{cm}	139	MPa
Flexural tensile strength	$f_{ctm,fl}$	18.2	MPa
Axial tensile strength	f_{ctm}	9.3	MPa
Density	ρ_c	2500	kg m^{-3}

friction, and statical inaccuracies have to be transferred between the single modules. These loads strongly depend on the individual parabolic trough system (e.g. size, number of modules, drive system) and must therefore be individually determined.

Solid Shell In a cooperation of seven research institutions and industrial partners (Table 4.2), a concrete collector was developed and built up in full scale. The project was led by the German Aerospace Center (DLR) and named "ConSol – Concrete Solar Collector" [42–44].

The collector, depicted in Figure 4.29, is composed of two modules that share a central drive. It can be divided into the superstructure, being the solid shell with the mirrors and sickles having gear wheels, and the substructure that consists of

Table 4.2 Collaborators of the project "ConSol."

Company/institution	Location (in Germany)
ALMECO GmbH	Bernburg
German Aerospace Center (DLR)	Cologne
Pfeifer Seil- und Hebetechnik GmbH	Memmingen
Ruhr University Bochum	Bochum
Solarlite CSP Technology GmbH	Duckwitz
Stanecker Betonfertigteilwerk GmbH	Borchen
TU Kaiserslautern	Kaiserslautern

Figure 4.29 Built-up and installed "ConSol" concrete collector at the precast fabrication "Stanecker." Picture: Sven Paustian.

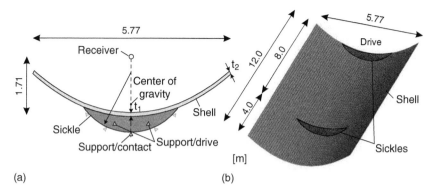

Figure 4.30 (a) Cross section of the shell with sickles and supports, (b) FE model.

a corresponding gear structure over an arch [52], footings, and strut-like interconnections. The movement of the collector occurs as a rolling of the superstructure relative to the supports. It takes place in between the sickles and the corresponding arch of the support. The gears prevent unintended sliding. Sickles and ground arch are shaped in such a way that the center of gravity of the moving superstructure remains on a horizontal level. Thus, no mechanical lifting work is needed to roll the superstructure in the daily course, except to overcome small impacts of friction and geometrical uncertainties. The sickles exhibit a circular shape, whereas the arches follow a cycloid function [53].

The form finding of the shell results from the minimization of the concrete material. The shell is characterized by two free values of the thicknesses, t_1 at the vertex and t_2 at the edge (Figure 4.30a), and a linear course in between. Tensile stresses are restricted to a share of the tensile strength, namely 6.0 MPa, and the deformations are controlled by a criterion of waviness, the so-called slope deviation [54]. A commercial FEM software package is used for the stress evaluations and superpositions from the single load cases. Figure 4.30b shows the FE model incl. the discretization and the supports.

In conclusion, t_1 results to 5.5 cm and t_2 to 3.5 cm yielding a remarkable small thickness of 4.5 cm in the average.

Bracing Pattern To elaborate the design, an optimal bracing layout is sought using one-material CTO. The idea is to locally stiffen the shell by cross-bracings and thus to reduce the shell thickness in between. Consequently, the overall material consumption decreases. For a conceptual design, three governing load cases are assumed and treated with uniform loadings q. They model dead loads in noon position (Figure 4.31, a1), one-sided wind loads (a2) and torsional effects from transferring drive loads from one shell to the next (a3). Figure 4.31 illustrates on its right the corresponding density distributions with a volume reduction factor of $\beta = 0.25$.

For robustness, the three load cases are now treated collectively. The objective function is enhanced to minimize the sum of compliance according to Section 4.1.6.

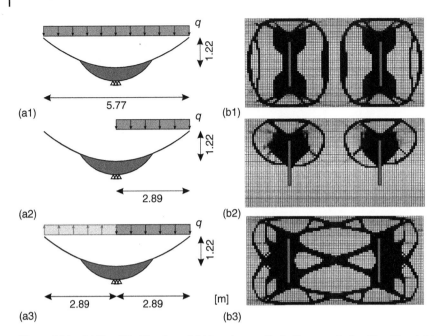

Figure 4.31 (a) Simplified loads and (b) topology optimization results for (1) self-weight, (2) wind load, and (3) torsional load for a parabolic shell collector.

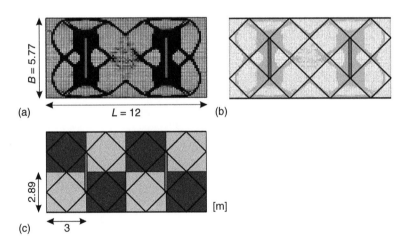

Figure 4.32 (a) Optimization result for multiple load cases, (b) pattern of stiffeners, and (c) possible formwork matrices.

For the sake of simplicity, all load cases (cf. Figure 4.31) are equally weighted and the volume fraction is held constant at $\beta = 0.25$.

Figure 4.32a shows the resulting density distribution derived from optimization. It is characterized by material accumulations lumped at the supports, especially at their outer corners, and almost linear courses of struts with a predominately diagonal orientation. In the next step, the density distribution is simplified to a purely

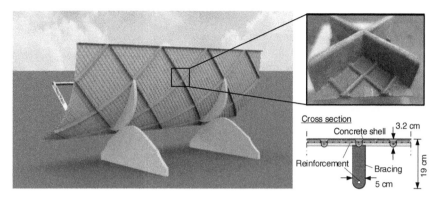

Figure 4.33 Visualization of the parabolic shell with stiffeners and a detail of the crossing bracings as a test specimen made from HPC.

uniform bracing pattern that exhibits crossing diagonals and longitudinal stiffeners at the outer borders (Figure 4.32b). Therefore, optimization acts as a visual design aid. The formwork of this pattern can be produced either as a hole or – taking advantage of the double symmetry and repetitions – by only two sub elements that repeat four times each (Figure 4.32c).

Figure 4.33 shows an implementation of the conceptual design into a RC structure. The bracings are shaped to inverted T-sections of constant height that support a thin shell. The shell itself picks up the ground pattern of a crosswise reinforcement mat that reinforces it. Doing so, a secondary bracing structure arises that orientates in the same diagonal direction as the primary bracings do. Overall, a residual concrete quantity of 2.3 cm smeared to an average shell thickness remains, so the initial average value of 4.5 cm (solid shell) is almost halved. As expected, the use of bracings located in the course of principal stresses yield much more efficient structures. This especially holds true for actions that provoke bending but to a similar extent for axial or membrane actions.

4.5.1.2 Heliostats

Point-focusing heliostats exhibit a three-dimensional shape of paraboloids and large focal lengths up to 1000 m [55]. Thus, their accuracy demands go far beyond the ones of parabolic troughs, as the distance to the receiver is considerably greater (cf. Figure 4.28b). The bending stiffness has to rise. Moreover, curvatures become small and collector surfaces get almost plane.

The conceptual design of heliostats made of concrete starts from the notion of bracings or grillage structures. Compared to a solid shell, dissolved grillages activate much larger bending stiffness with the same amount of material. Mirrors locate on top to span in between the single struts of the grillage.

Moreover, heliostat exhibit rotations around two axis, namely around the vertical one (azimuth) and one horizontal axis (elevation) to achieve suitable orientations for sun tracking. Considering these biaxial movements, a central hinge makes sense. It is placed in the center of gravity of the heliostat (superstructure) and connects it to the vertical column of the substructure. Thus, sun tracking occurs without lifting

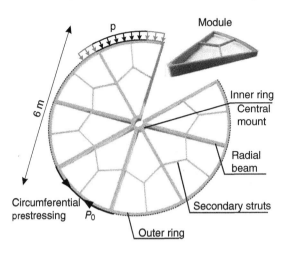

Figure 4.34 Conceptual design of the strut-like, modular heliostat made of HPC.

works despite negligible contributions from friction or unintended geometric inaccuracies. Figure 4.34 shows the conceptual idea of the concrete heliostat. Its major features are

- A circular outer shape so that the distance to the central hinge (inside the inner ring) for rotations is balanced for all directions [56]. The aperture area yields about 30 m², what is common for cost-optimized heliostats [57].
- A surrounding prestressing P_0 by a tendon that activates deviation forces $p = P_0/R$ to overpress tensile stresses from bending in the radial struts and to activate the high compressibility of concrete. R denotes the radius of 3 m.
- Struts in radial and circumferential direction to provide bending stiffness, regular supports for the mirrors and an outer concrete ring to continuously fix and deviate the tendon.
- A subdivision of the heliostat into uniform modules. The modules shape like "pieces of cake" defined by their opening angle φ and hold together by just dry joints between the radial beams. The joints are compressed in ring direction by the outer prestressing.
- The HPC embodied in Table 4.1 is used. Here, its high Young's modulus as well as its high compressive strength are the relevant parameters. Tensile concrete stresses play a subordinate role as they are neutralized to the most extent by prestressing.

Form Finding The form finding is performed in two steps. First, a general layout of the radial stiffness demand is derived. This step aims for a pre-dimensioning of the radial beams. The heliostat is treated as rotationally symmetric and oriented in an upward position under dead loads and wind. Second, the bracing structure and its dissolution into modules are treated using one-material CTO.

The radial bending stiffness is estimated from the simple static system of a cantilever. The transfer to such a replacement structure of a cantilever is done with symmetry reduction methods [58, 59], since both loads as well as geometry can be idealized as rotationally symmetric. Figure 4.35a shows the underlying concept of

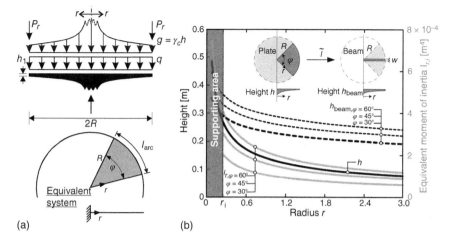

Figure 4.35 (a) Rotationally symmetric plate with hyperbolical thickness and loads according to [58] with segmental structure and equivalent statistical system of a cantilever arm, (b) thickness of plate (black line) with corresponding moment of inertia I_r (grey line) and equivalent height of beams h_{beam} (dotted line) for different opening angles φ.

a circular plate with a hyperbolic thickness under uniform and line-like loadings. The loads are defined by the distributed dead load g from concrete, a uniform load $q = 0.5\,\text{kN/m}^2$ compromising wind and mirror elements, and an outer line-like load $P_r = 0.25\,\text{kN/m}$ representing the dead load of the surrounding ring. The thickness of the plate is given to

$$h = h_1\left(\frac{r}{R}\right)^{-1/2} \tag{4.64}$$

with h_1 defining the thickness at the edge of the plate. R denotes the radius and r the radial coordinate, respectively. Prestressing is not yet considered, because it does not significantly contribute to the stiffness. It should be noted that the applied wind loads just serve for a pre-design. They pronouncedly vary in their distribution and extent in dependence upon the actual orientation of a heliostat (cp. e.g. [60, 61]).

The necessary stiffness results from limiting the vertical deformation of the outer ring. There, the extreme rotations occur that deform the paraboloid and thus disturb the sun focus. The stiffness is derived from the radial moment of inertia I_r

$$I_r = \frac{1}{12}Rh_1^3\varphi\sqrt{r/R} \tag{4.65}$$

that develops over r in such a way that all cross sections reasonably contribute to the outer deflection. h_1 denotes an initial height at the outer ring.

Figure 4.35b shows the resulting courses of the height h (left) and the moment of inertia I_r (right) as functions over the radial location r. The values are given for one resulting strut per module and a strut width of 10 cm. So, they depend on the opening angle φ and lump necessary stiffness to just one beam per module. Thus, the two beams per module exhibit halve of the value each. Both h and I_r follow a hyperbolic course that cuts of at the inner ring r_i, where the interconnection to the substructure is placed.

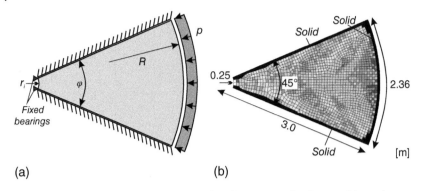

Figure 4.36 (a) Segmental module with line-like prestressing force p, (b) topology optimization result for a module with an opening angle of 45°.

The second step of form finding attributes the identification of most suitable pattern of bracings. One-material CTO and a commercial software are used for this. The derivation assumes the governing load case of prestressing that provokes line-like, inward forces p. It refers to just one module described by its opening angle, the inner and outer radius, the loading as well as lateral and inner bearings. Figure 4.36 illustrates the static system (a) and an example of discretization into shell elements (b).

As expected, CTO under the distributed load p as well as line-like bearings does not yield clear 0-1 material distributions but rather smeared demands of densities. They hardly aid to identify bracings. Figure 4.36b shows such a density distribution by gray scales. The volume fraction is set to $\beta = 0.30$.

As a remedy, the line load p is lumped to resultants and applied by single loads P. Moreover, the lateral line supports are replaced by single bearings. They clearly steer the locations of contact between the modules and directly report reliable local contact forces that are necessary to insure sufficient compression for the overall integrity of the single modules to act as a hole. In realizations, a local thin layer of concrete is shaped that builds this local dry surface-to-surface contact.

Figure 4.37 shows three design studies varying the opening angle ($\varphi = 30°$, 45° or 60°), locations and number of forces (two or four) as well as locations of lateral bearings (close to the outer ring, close to the inner ring, centered). Now, clear 0-1 density distributions arise that help to identify the most relevant traces of force transfer (center). They can be easily transferred into bracing pattern (right) that include the outer ring – necessary for the tendon – lateral struts with local contact points to adjacent modules and intermediate struts to stiffen between the lateral contacts. All struts also support the mirrors that need a regular grid of supports to keep local free spans small.

Prototype A true-scale prototype of a concrete heliostat has been developed and realized within a "knowledge transfer" project [38, 62]. The heliostat exhibits an inner radius r_i of 0.15 m, an outer radius R of 1.60 m and a mirror area of about 8 m² [62]. Figure 4.38 shows the super- and the substructure in front and back view.

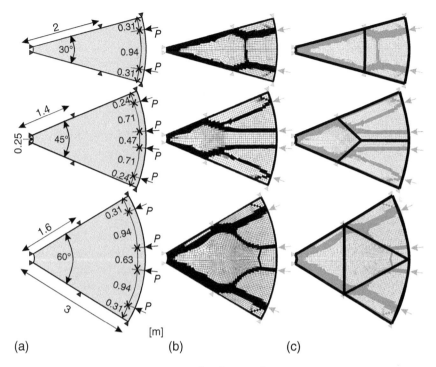

Figure 4.37 Design studies for $\varphi = 30°, 45°$ and $60°$: (a) Design space with loads and bearings, (b) topology optimization results, (c) identified strut pattern.

Figure 4.38 Prototype of a modular concrete heliostat at the DLR solar tower in Jülich, Germany. Pictures: Patrick Forman, Marius Schellen.

Table 4.3 Collaborators of the heliostat "transfer" project

Company/institution	Location (in Germany)
ALMECO GmbH	Bernburg
Durcrete GmbH	Limburg
German Aerospace Center (DLR)	Stuttgart/Jülich
Innogration GmbH	Bernkastel-Kues
Ruhr University Bochum	Bochum
TU Kaiserslautern	Kaiserslautern

The heliostat consists of four modules with an opening angle of $\varphi = 90°$ and a bracing pattern similar to the second proposal (Y-shaped inner struts) of Figure 4.37. The weight of the concrete sums up to 340 kg in total what corresponds to a material usage of about 1.7 cm concrete relative to the mirror surface. The heliostat is produced in the laboratory of TU Kaiserslautern, Germany, using a specific 3D-cutout polystyrene formwork. It is connected to a steel column and a concrete footing via an inner steel ring fixed to a horizontal steel tube that runs on ball bearings. Two linear drives account for the rotations. The mirror surface follows the subdivision into eight "pieces of cake" and provides the notion of a blossom of a flower.

The participating research institutes as well as industrial partners are summarized in Table 4.3.

4.5.2 Ultra-light Beams

Using the bi-material topology optimization approach from Section 4.3, a simple RC beam is optimized in [6, 63–65]. The aim is to reduce its structure to the essential load-bearing design and thereby demonstrate the vast savings potential in material and GHG emissions. However, both load-bearing capacity and stiffness must not be compromised.

The RC beam serves as reference and is loaded in four-point bending. It employs a solid rectangular cross section with $B/H = 150/300$ mm, a span width of 2400 mm, and a total length of 2800 mm (Figure 4.39a). The materials used are, on the one hand, NC with mean compressive strength $f_{cm} = 38$ MPa (type C30/37 according to Eurocode 2 [66]) and, on the other hand, reinforcing steel exhibiting a projected yield strength of $f_y = 500$ MPa (type B500 according to the German standard DIN 488 [67]). The beam is designed in such a way that its bending load-bearing capacity is fully utilized. In doing so, the lower tensile reinforcement yields while the compression zone crushes at the same time. The embedded stirrups prevent premature shear force failure. Both the maximum load of 205.3 kN obtained in the experiment as well as its load–deflection curve (gray solid line in Figure 4.40) representing the structural stiffness serve as reference for the subsequent optimization.

The bi-material topology optimization is applied to the RC beam. To counteract the expected loss of stiffness due to the targeted material reduction, the design space height is enlarged to 400 mm (Figure 4.39b). The supports are defined at

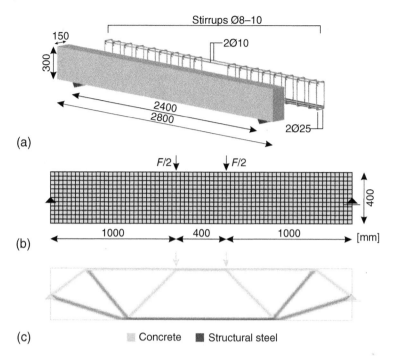

(a)

(b)

(c)

Concrete ■ Structural steel

Figure 4.39 (a) Reference RC beam, (b) design space and FE mesh, (c) bi-material optimization result, according to [63].

medium height to allow for better shape adaptation to the bending moment. For the FE mesh, $240 \times 40 = 9600$ quadrilateral elements are employed. The applied loads amount to the ultimate load reached by the reference beam in the experiment, namely $F = 205.3$ kN. The material parameters of the compression-only material are $E_c = 48\ 000$ MPa, and $v_c = 0.20$. They correspond to an ultra-high performance concrete (UHPC), hence, compared to the reference beam, a material update is aimed for in order to exploit a maximum possible weight reduction potential. On the other hand, the material parameters of the tension-only material are set to $E_s = 200\ 000$ MPa and $v_s = 0.3$, thus representing structural steel.

Figure 4.39c shows the optimization result for a residual volume of 10 % ($\beta = 0.10$). Material redistribution within the design space leads to a truss-like structure composed of predominantly axially loaded compression and tension struts, which are interconnected in nodes. The compression struts consist of UHPC and the tension struts of mild steel. Bending moment and shear force are virtually eliminated. Two alternative designs with gradually reduced weight are derived from this design proposal: a RC truss structure and a hybrid concrete–steel truss structure, cf. Figure 4.40a.

The RC truss structure is designed and fabricated using NC and reinforcing steel in conventional reinforced concrete design. Its weight is reduced by 53 % from 305 kg to 143 kg compared to the reference beam. At the same time, the material-induced GHG emissions decrease by 40 % [6, 63]. However, the load-bearing capacity is maintained and the stiffness increases considerably (dashed line in Figure 4.40b).

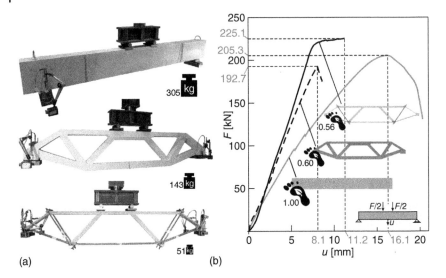

Figure 4.40 (a) Reference beam, RC truss structure and hybrid truss structure in the test setup, (b) related results of the load test and relative material-induced GHG footprint, according to [6].

The hybrid truss structure goes one step beyond. Regarding the compressive struts, in this case they are made of UHPC exhibiting a mean compressive strength of f_{cm} = 145 MPa. To increase ductility, micro steel fibers are also added. Stainless steel meshes with wire diameters of 2 mm serve as structural reinforcement, since conventional rebars hardly fit within the very slender cross sections. The tensile struts, on the other hand, are released from any concrete including the concrete cover and consist only of sectional steel. The latter exhibits a yield strength of f_y = 355 MPa (type S355 according to Eurocode 3 [68]). These further design adjustments and material updates allow vast weight reductions of up to 83 % to only 51 kg compared to the reference beam. Concurrently, GHG emissions drop by 44 % [6, 63]. In doing so, the load-bearing capacity of the hybrid truss structure is not only preserved but can even be improved by 10 % (225.1 kN). Its stiffness resembles that of the RC truss structure and is thus also much higher compared to that of the reference beam.

The stepwise conversion of a conventional RC beam into an ultra-light hybrid truss structure reveals the vast potential of optimization-aided design approaches in terms of resource savings and reduction of GHG emissions without requiring compromises in performance.

References

1 Hong, J., Shen, G.Q., Feng, Y. et al. (2015). Greenhouse gas emissions during the construction phase of a building: a case study in China. *Journal of Cleaner Production* 103: 249–259. ISSN: 09596526.

2 Mehta, P.K. (2001). Reducing the environmental impact of concrete. *Concrete International* 23 (10): 61–66.

3 Rodrigues, F.A. and Joékes, I. (2011). Cement industry: sustainability, challenges and perspectives. *Environmental Chemistry Letters* 9 (2): 151–166. https://doi.org/10.1007/s10311-010-0302-2.

4 Suhendro, B. (2014). Toward green concrete for better sustainable environment. *Procedia Engineering* 95: 305–320. https://doi.org/10.1016/j.proeng.2014.12.190.

5 Worrell, E., Price, L., Martin, N. et al. (2001). Carbon dioxide emissions from the global cement industry. *Annual Review of Energy and the Environment* 26 (1): 303–329. https://doi.org/10.1146/annurev.energy.26.1.303.

6 Gaganelis, G. (2020). Ultra-light hybrid concrete-steel beams. PhD thesis. Bochum: Ruhr University Bochum.

7 Forman, P., Gaganelis, G., and Mark, P. (2019). Optimierungsgestütztes Entwerfen und Bemessen. In: *Realität – Modellierung – Tragwerksplanung. 23. Dresdner Baustatikseminar* (eds. M. Kaliske and W. Graf), TU Dresden, 5–23.

8 Gaganelis, G., Forman, P., and Mark, P. (2021). Stahlbeton optimiert - für ein Mehr an Weniger. In: *Nachhaltigkeit, Ressourceneffizienz und Klimaschutz* (ed. B. Hauke, Institut Bauen und Umwelt e.V., and DGNB e.V.). Ernst & Sohn, 159–167. ISBN: 9783433033357.

9 Bendsøe, M.P. and Sigmund, O. (1999). Material interpolation schemes in topology optimization. *Archive of Applied Mechanics* 69 (9–10): 635–654. https://doi.org/10.1007/s004190050248.

10 Bendsøe, M.P. (1989). Optimal shape design as a material distribution problem. *Structural Optimization* 1 (4): 193–202. https://doi.org/10.1007/BF01650949.

11 Bendsøe, M.P. and Sigmund, O. (2004). *Topology Optimization: Theory, Methods, and Applications*. Berlin: Springer-Verlag.

12 Mlejnek, H.-P. and Schirrmacher, R. (1993). An engineer's approach to optimal material distribution and shape finding. *Computer Methods in Applied Mechanics and Engineering* 106 (1–2): 1–26. https://doi.org/10.1016/0045-7825(93)90182-W.

13 Rozvany, G., Zhou, M., and Birker, T. (1992). Generalized shape optimization without homogenization. *Structural Optimization* 4 (3–4): 250–252. https://doi.org/10.1007/BF01742754.

14 Zhou, M. and Rozvany, G. (1991). The COC algorithm, Part II: Topological, geometrical and generalized shape optimization. *Computer Methods in Applied Mechanics and Engineering* 89 (1–3): 309–336. https://doi.org/10.1016/0045-7825(91)90046-9.

15 Sigmund, O. and Petersson, J. (1998). Numerical instabilities in topology optimization: a survey on procedures dealing with checkerboards, mesh-dependencies and local minima. *Structural Optimization* 16 (1): 68–75. https://doi.org/10.1007/BF01214002.

16 Sigmund, O. (2007). Morphology-based black and white filters for topology optimization. *Structural and Multidisciplinary Optimization* 33 (4–5): 401–424. https://doi.org/10.1007/s00158-006-0087-x.

17 Wang, F., Lazarov, B.S., and Sigmund, O. (2011). On projection methods, convergence and robust formulations in topology optimization. *Structural and Multidisciplinary Optimization* 43 (6): 767–784. ISSN: 1615-147X.

18 Sigmund, O. (1997). On the design of compliant mechanisms using topology optimization. *Mechanics of Structures and Machines* 25 (4): 493–524. https://doi.org/10.1080/08905459708945415.

19 Smarslik, M. (2019). Optimization-based design of structural concrete using hybrid reinforcements. PhD thesis. Bochum: Ruhr University Bochum.

20 Arora, J.S. (2011). *Introduction to Optimum Design*. Amsterdam: Elsevier Academic Press.

21 Harzheim, L. (2008). *Strukturoptimierung: Grundlagen und Anwendungen*. Frankfurt am Main: Harri Deutsch.

22 Smarslik, M., Ahrens, M.A., and Mark, P. (2019). Toward holistic tension- or compression-biased structural designs using topology optimization. *Engineering Structures* 199. https://doi.org/10.1016/j.engstruct.2019.109632.

23 Haftka, R.T. and Gürdal, Z. (1993). *Elements of Structural Optimization*. Dordrecht: Kluwer.

24 Gaganelis, G., Jantos, D., Mark, P., and Junker, P. (2019). Tension/compression anisotropy enhanced topology design. *Structural and Multidisciplinary Optimization* 59 (6): 2227–2255.

25 Sigmund, O. (2001). Design of multiphysics actuators using topology optimization – Part II: Two-material structures. *Computer Methods in Applied Mechanics and Engineering* 190 (49–50): 6605–6627. https://doi.org/10.1016/S0045-7825(01)00252-3.

26 Sigmund, O. and Torquato, S. (1997). Design of materials with extreme thermal expansion using a three-phase topology optimization method. *Journal of the Mechanics and Physics of Solids* 45 (6): 1037–1067. https://doi.org/10.1016/S0022-5096(96)00114-7.

27 Bogomolny, M. and Amir, O. (2012). Conceptual design of reinforced concrete structures using topology optimization with elastoplastic material modeling. *International Journal for Numerical Methods in Engineering* 90 (13): 1578–1597. https://doi.org/10.1002/nme.4253.

28 Cai, K. (2011). A simple approach to find optimal topology of a continuum with tension-only or compression-only material. *Structural and Multidisciplinary Optimization* 43 (6): 827–835. https://doi.org/10.1007/s00158-010-0614-7.

29 Kirsch, U. (1990). On singular topologies in optimum structural design. *Structural and Multidisciplinary Optimization* 2 (3): 133–142. https://doi.org/10.1007/BF01836562.

30 Rozvany, G. (2001). On design-dependent constraints and singular topologies. *Structural and Multidisciplinary Optimization* 21 (2): 164–172. https://doi.org/10.1007/s001580050181.

31 Andreassen, E., Clausen, A., Schevenels, M. et al. (2011). Efficient topology optimization in MATLAB using 88 lines of code. *Structural and Multidisciplinary Optimization* 43 (1): 1–16.

32 Smarslik, M., Kämper, C., Forman, P. et al. (2016). Topologische Optimierung von Betonstrukturen. In: *Festschrift für Manfred Curbach* (eds. S. Scheerer, U. van Stipriaan, and W. Leiberg), 234–255. Dresden: Technische Universität Dresden.

33 Karoumi, R., Andersson, A., and Sundquist, H. (2006). Static and dynamic load testing of the new svinesund arch bridge. *The International Conference on Bridge Engineering*.

34 Kalogirou, S.A. (2004). Solar thermal collectors and applications. *Progress in Energy and Combustion Science* 30 (3): 231–295. https://doi.org/10.1016/j.pecs .2004.02.001.

35 Evangelisti, L., de Lieto Vollaro, R., and Asdrubali, F. (2019). Latest advances on solar thermal collectors: a comprehensive review. *Renewable and Sustainable Energy Reviews* 114: 1085–1091. https://doi.org/10.1016/j.rser.2019.109318.

36 Jebasingh, V.K. and Herbert, G.M.J. (2016). A review of solar parabolic trough collector. *Renewable and Sustainable Energy Reviews* 54: 1085–1091. https://doi .org/10.1016/j.rser.2015.10.043.

37 Fuqiang, W., Ziming, C., Jianyu, T. et al. (2017). Progress in concentrated solar power technology with parabolic trough collector system: a comprehensive review. *Renewable and Sustainable Energy Reviews* 79: 1314–1328. https://doi.org/ 10.1016/j.rser.2017.05.174.

38 Forman, P., Penkert, S., Kämper, C. et al. (2020). A survey of solar concrete shell collectors for parabolic troughs. *Renewable and Sustainable Energy Reviews* 134: 110331. https://doi.org/10.1016/j.rser.2020.110331.

39 Pfahl, A. (2014). Survey of heliostat concepts for cost reduction. *Journal of Solar Energy Engineering* 136 (1). https://doi.org/10.1115/1.4024243.

40 Pfahl, A., Coventry, J., Röger, M. et al. (2017). Progress in heliostat development. *Solar Energy* 152: 3–37. https://doi.org/10.1016/j.solener.2017.03.029.

41 Weinrebe, G. and Balz, M. (2019). Efficient steel structures for solar thermal power plants. *Stahlbau* 88 (6): 529–536. https://doi.org/10.1002/stab.201900037.

42 Forman, P., Stallmann, T., Mark, P., and Schnell, J. (2018). Multi-level optimisation of parabolic shells with stiffeners made from high-performance concrete. In: *High Tech Concrete: Where Technology and Engineering Meet* (eds. D.A. Hordijk and M. Lukovic), 2503–2511. Cham: Springer International Publishing. ISBN: 978-3-319-59470-5. https://doi.org/10.1007/978-3-319-59471-2_285.

43 Penkert, S., Forman, P., Mark, P., and Schnell, J. (2019). Conceptual design and construction of an original scale parabolic trough collector made of high-performance concrete. *Beton- und Stahlbetonbau* 114 (11): 806–816.

44 Krüger, D., Penkert, S., Schnell, J. et al. (2019). Development of a concrete parabolic trough collector. *SolarPACES 2018: International Conference on Concentrating Solar Power and Chemical Energy Systems*. https://doi.org/10.1063/1 .5117626.

45 Forman, P., Penkert, S., Mark, P., and Schnell, J. (2020). Design of modular concrete heliostats using symmetry reduction methods. *Civil Engineering Design* 2 (4): 92–103. https://doi.org/10.1002/cend.202000013.

46 Sagmeister, B. (2017). *Maschinenteile aus zementgebundenem Beton*. Berlin: Beuth.

47 Lüpfert, E., Geyer, M., Schiel, W. et al. (2001). EuroTrough design issues and prototype testing at PSA. *Campbell (Hg.) 2001 – Proceedings of the ASME international*.

48 Schiel, W. (2012). Kollektorentwicklung für solare Parabolrinnenkraftwerke. *Bautechnik* 89 (3): 182–191.

49 Müller, S., Forman, P., Schnell, J., and Mark, P. (2013). Lightweight shells made of high–strength concrete for parabolic troughs in solar power plants – conceptual design and creation of a prototype. *Beton- und Stahlbetonbau* 108 (11): 752–762.

50 Forman, P., Kämper, C., Stallmann, T. et al. (2016). Parabolic shells made from high–performance concrete for solar collectors. *Beton- und Stahlbetonbau* 111 (12): 851–861. https://doi.org/10.1002/best.201600051.

51 Winkelmann, U., Kämper, C., Höffer, R. et al. (2020). Wind actions on large-aperture parabolic trough solar collectors: Wind tunnel tests and structural analysis. *Renewable Energy* 146: 2390–2407.

52 Penkert, S. (2020). Zur Auslegung von Betonverzahnungen für Parabolrinnen unter Berücksichtigung des tribologischen Verhaltens von Hochleistungsbeton. PhD thesis. Kaiserslautern: TU Kaiserslautern.

53 Weißbach, R. (2012) Die abrollende Parabolspiegelrinne. Schutzrecht DE102011011805.

54 Krüger, D., Pandian, Y., Hennecke, K., and Schmitz, M. (2008). Parabolic trough collector testing in the frame of the REACt project. *Desalination* 220 (1–3): 612–618. https://doi.org/10.1016/j.desal.2007.04.062.

55 Röger, M., Prahl, C., and Ulmer, S. (2008) Fast Determination of Heliostat Shape and Orientation by Edge Detection and Photogrammetry. In: 14th CSP SolarPACES Symposium 2008, Las Vegas, USA.

56 Keck, T., Balz, M., Göcke, V. et al. (2018). Hami – The first Stellio solar field. *SolarPACES 2018: International Conference on Concentrating Solar Power and Chemical Energy Systems*.

57 Cordes, S., Prosinecki, T.C., and Wieghardt, K. (2012). An approach to competitive heliostat fields. *18th SolarPACES Conference*, Marrakech, Marocco.

58 Márkus, G. and Otto, J. (1978). *Theorie und Berechnung rotationssymmetrischer Bauwerke*. Düsseldorf: Werner. ISBN: 3804126510.

59 Bocklenberg, L. and Mark, P. (2020). Thick slab punching with symmetry reductions. *Structural Concrete* 21 (3): 875–889. https://doi.org/10.1002/suco.201900480.

60 Gong, B., Wang, Z., Li, Z. et al. (2013). Fluctuating wind pressure characteristics of heliostats. *Renewable Energy* 50: 307–316.

61 Peterka, J.A. and Deickson, R. (1992). Wind Load Design Methods for Ground-Based Heliostats and Parabolic Dish Collectors. National Technical Information Service, US Department of Commerce, Springfield, VA, USA. *Technical Report SAND92-7009*.

62 Forman, P., Penkert, S., Mark, P., and Schnell, J. (2021). Heliostate aus UHPC – modularer Ultraleichtbau. *BFT International* 87 (2): 40–41.

63 Gaganelis, G. and Mark, P. (2019). Downsizing weight while upsizing efficiency: An experimental approach to develop optimized ultra–light UHPC hybrid beams. *Structural Concrete* 20 (6): 1883–1895.

64 Gaganelis, G. and Mark, P. (2019). Ultra-light beams as concrete-steel hybrid. In: *Concrete - Innovations in Materials, Design and Structures, FIB Symposium 2019*. Fédération internationale du béton, fib (ed. W. Derkowski et al., 1114–1120. Krakow.

65 Mark, P. and Gaganelis, G. (2019). Ultra-light beams as concrete-steel hybrid. *BFT International* 85 (2): 95.

66 EN 1992-1-1 (2014). *Eurocode 2: Design of Concrete Structures – Part 1-1: General Rules and Rules for Buildings*. Brussels: Comité Européen de Normalisation (CEN).

67 DIN 488-1 (2009). *Reinforcing Steels – Part 1: Grades, Properties, Marking*. Berlin: Beuth.

68 EN 1993-1-1 (2010). *Eurocode 3: Design of Steel Structures – Part 1-1: General Rules and Rules for Buildings*. Brussels: Comité Européen de Normalisation (CEN).

5

Internal Force Flow

Key learnings after reading this chapter:

- What are practical optimization methods for identifying the internal force flow of structures?
- How can the results being used to develop robust strut-and-tie models?
- Which aspects have to be taken into account when applying the techniques in practice?

Discontinuity regions (D-regions), meaning structural parts for which no plane strain state can be assumed due to their geometry or load condition, have to be designed with strut-and-tie models (STMs). Although standard models are available for a variety of D-regions, however, components for which individual STMs must be developed are not unusual. Beams, walls, and ceilings with random openings are the most common recurring structural elements in this context.

Usually, the procedure for setting up a STMs is iterative. After computing the stress state via FE analysis, the principal stresses are merged into struts, which in turn must represent a STMs that is in static equilibrium. This requires frequent corrections to the strut configuration, which is very time-consuming. Moreover, the finally identified STMs leaves uncertainty regarding the equilibrium and whether better models exists, which either require less reinforcement or reflect the stress state more accurately.

This chapter introduces structural optimization methods, which allow to automatically compute optimized STMs in order to derive reinforcement layouts (Figure 5.1). Recommendations for an easy practical application are given. The goal is to generate customized STMs without much effort and within a reasonable amount of time.

Optimization Aided Design: Reinforced Concrete, First Edition.
Georgios Gaganelis, Peter Mark, and Patrick Forman.
© 2022 Ernst & Sohn GmbH & Co.KG. Published 2022 by Ernst & Sohn GmbH & Co.KG.

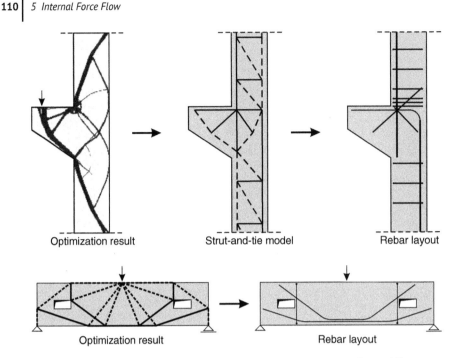

Optimization result   Strut-and-tie model   Rebar layout

Optimization result   Rebar layout

Figure 5.1 Overview of Chapter 5: strut-and-tie models, partly according to [1].

5.1 Preliminaries

According to [2], the best strut-and-tie model (STM) among a variety of possible variants is the one with the lowest sum over the axial forces (N_i) multiplied by the associated deformations ($L_i\varepsilon_i$) (cf. Section 2.5.2) over all members $i \in [1, m]$:

$$\sum_{i=1}^{m} N_i L_i \varepsilon_i \to \min \tag{5.1}$$

where L_i is the member length, ε_i is its strain, and m is the number of members. In this context, it should be recalled that Eq. (5.1) equals the *principle of minimum strain energy* for linear elastic material behavior. On this basis, the well-known attempt of minimizing the mean structural compliance c in structural optimization becomes the ideal objective function of an optimization problem addressing STMs. The interrelation between the principle of minimum strain energy and the mean structural compliance results from the fact that for a body behaving linear-elastically, the internal work (W_i) just corresponds to the external work (W_a):

$$\underbrace{\frac{1}{2}\int_{\Omega} \sigma\varepsilon\, dV}_{W_i} = \underbrace{\frac{1}{2}\int_{u} F\, du}_{W_a} \tag{5.2}$$

where σ and ε are the stresses and strains, respectively. V is the volume of the body, u are the displacements, and F are the external loads. When using the FE method for discretization, the relation is given as:

$$\frac{1}{2}\sum_{i=1}^{m}L_i\varepsilon_i = \underbrace{\frac{1}{2}\sum_{i=1}^{m}F_iu_i}_{c} \tag{5.3}$$

Eventually, it can be seen that the right-hand side of Eq. (5.3) equals half the value of c.

In the following, optimization-aided approaches for generating STMs are presented. In addition to time savings, it is also find that they may provide also more cost-efficient reinforcement layouts [3].

5.2 Continuum Topology Optimization (CTO) Approach

Related Examples: 5.1–5.5.

A simple way to develop STMs is to first identify the internal force flow using the classical continuum-based minimum compliance with volume constraint topology optimization (CTO), cf. Section 4.1, and then convert the resulting material distribution into a STMs. Modeling of the design space should include possible boundary conditions of the adjoining structural areas as specified in the recommendations given in Section 5.3.4. Subsequently, the compression and tension struts need to be determined and their forces must be computed. Hence, this procedure requires manual post-processing after the optimization results have been calculated. However, depending on the material distribution, in some cases, several STMs are conceivable. Furthermore, it can happen that STMs derived directly from CTO results become kinematically intedeterminate and thus have to be adapted "by hand" in order to ensure that computing the member forces is possible. Despite these limitations, the approach is very intuitive and relies on a simple but stable optimization approach that can be applied reliably and quickly to a wide range of problems.

The complexity of the optimization results and consequently of the ensuing STMs can be controlled in two ways. The first is the number of finite elements (N_e) used to discretize the design space. However, if numerical regularization is applied, for example, the filter methods presented in Section 4.1.3, mesh-independent results can be achieved in spite of different mesh sizes. It is therefore recommended to prefer finer meshes with many elements in order to enable more detailed structural designs to be represented in the optimization results. If desired, the level of detail can then be reduced by recalculating the results with increased filter radii (Figure 5.2). The second parameter to control complexity is the volume fraction $\beta \in [0,1]$. It constrains the available material amount relative to that of the initial design space. Rather lower volume fractions $\beta \leq 0.5$ should be preferred in order to identify distinct struts from the computed optimized material distribution, otherwise the transformation into STMs could become cumbersome due to overloaded design spaces.

The recommended procedure is illustrated in Figure 5.2. Starting with the generated FE model, material distributions for different values of β are calculated using

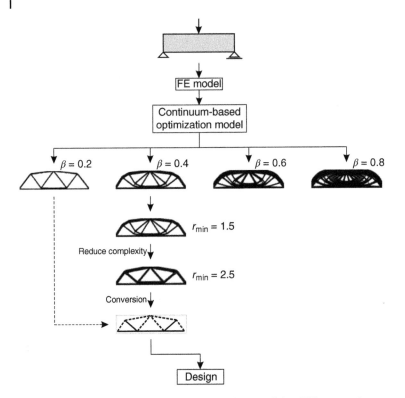

Figure 5.2 Recommended procedure for the STMs striving CTO approach.

a low filter radius. The most applicable result is then chosen, whose complexity can be further decreased, if desired, by increasing the filter radius and recalculating the material distribution. Thereafter, the struts are identified, a related STMs is generated, and its member forces are computed.

5.3 Truss Topology Optimization (TTO) Approach

Related Examples: 5.6–5.13.

5.3.1 Problem Statement

Since STMs consist of axially loaded struts or ties connected in nodes, alternatively, truss topology optimization (TTO) can be used to compute reasonable models. The main advantage over the continuum-based approach from Section 5.2 is obvious: the post-processing step of transferring the material distribution into a truss structure and calculating its member forces is omitted because the STMs is obtained directly as the solution to the optimization problem. Consequently, the member forces and thus the required reinforcement amounts are provided instantaneous. A further benefit is that quantitative information regarding the number of active trusses of the STMs is available which can be used to directly control the extent of complexity.

When using TTO, each bar equals one finite element. The feasible design space is defined using a mesh of geometric elements whose vertices coincide with the finite element nodes. This fundamental geometrical division of the design space is referred to as the *base mesh*. The nodes are connected to each other to a predefined extent (the so-called *level of connectivity*, *Lvl*) by the finite elements, that is the trusses, and thus form the initial structure referred to as the *ground structure* (Figure 5.3) [4–8]. With *Lvl* = 1, each node is connected to its neighbor node, *Lvl* = 2 additionally involves connections up to second-degree neighbors, *Lvl* = 3 leads to connections up to third-degree neighbors, and so on. Using the example of a simple beam loaded by a point load, Figure 5.4 depicts ground structures for different base meshes and levels of connectivity.

The FE model (analysis model) is linked to the optimization problem using the design variables x_e where $e \in [1, N_e]$ and N_e is the number of finite elements, which is equal to the number of bars in the ground structure. The design variables can approximately be interpreted as normalized cross-sectional areas of the bars:

$$x_e \approx \frac{A_e}{A_{\max}} \tag{5.4}$$

where A_e is the cross-sectional area of member e and $A_{\max}$ is the upper bound of the cross sections. Thereby, x_e takes values within the set [0,1]. For numerical reasons in order to prevent a singular stiffness matrix, the lower bound of the design variables

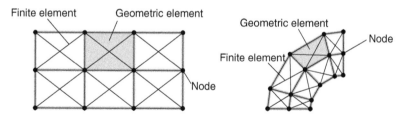

Figure 5.3 Examples for the components of a ground structure: geometric elements, finite elements, and nodes.

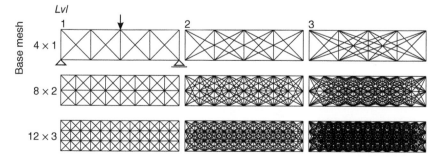

Figure 5.4 Different ground structures for a simple beam.

must actually be a small nonzero number. To overcome this, the subsequent interpolation approach can be used alternatively to keep the boundaries unchanged [9]:

$$A_e(x_e) = \left[(1 - x_{min}) x_e + x_{min} \right] A_{max} \tag{5.5}$$

In Eq. (5.5), x_{min} is a nonzero number that should be chosen small enough to ensure
that quasi-zero bars exert a negligible influence on the structure, for example,
$x_{min} = 10^{-6}$. The analysis model and the optimization model are linked together
using the so-called SIMP (Solid Isotropic Material with Penalization) approach
[10–15], whereby no penalty exponent is applied here:

$$\mathbf{k}_e(x_e) = A_e(x_e) L_e \mathbf{k}_e^0 \tag{5.6}$$

where $\mathbf{k}_e$ is the element stiffness matrix in global coordinates, L_e is the member
length, and $\mathbf{k}_e^0$ is the unit stiffness matrix in global coordinates. The latter is defined
for plane problems as follows:

$$\mathbf{k}_e^0 = \frac{E}{L_e^2} \begin{bmatrix} l^2 & lm & -l^2 & -lm \\ lm & m^2 & -lm & -m^2 \\ -l^2 & -lm & l^2 & lm \\ -lm & -m^2 & lm & m^2 \end{bmatrix} \tag{5.7}$$

where $l = \cos(\alpha)$ and $m = \sin(\alpha)$ transform a bar rotated by the angle α to the global
coordinate system (x, y), cf. Figure 5.5. The extension to spatial problems is straightforward, see for example [16].

As already outlined in Section 5.1, minimizing the mean structural compliance
c (Eq. (5.3)) is a suitable objective function for computing STMs. In doing so, the
maximum stiffness design is sought as optimization result. In order to prevent trivial
solutions, i.e. assigning A_{max} to each member, a constraint on the overall volume can
be applied, which limits the available material amount to be distributed within the
design space:

$$V \leq \beta V^0 \tag{5.8}$$

Here, V is the structure's volume, $\beta \in [0,1]$ is the volume fraction, and V^0 is the
initial volume of the ground structure.

However, one problem with this type of formulation is that the results highly
depend on the base mesh and the level of connectivity (*Lvl*). Coarse base meshes
and small values for *Lvl* restrict the solution space and generally lead to more
simple structures with fewer members. With finer base meshes and higher values
of *Lvl*, the solution space expands, but the structures become increasingly complex

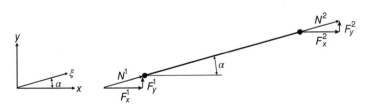

Figure 5.5 Transformation of a bar from local (ξ) to global (x, y) coordinates.

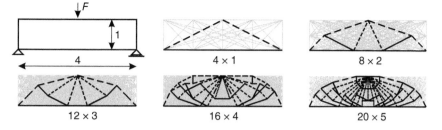

Figure 5.6 Dependence of the results on the ground structure (here: $Lvl = 4$).

and consist of a large number of struts, which makes them ineffective for STMs and thus impractical for placing and dimensioning reinforcements (Figure 5.6). For this reason, it is desirable to control the complexity of the optimization results even for elaborate ground structures in order to allow computing reasonable reinforcement layouts.

Only a few papers deal with the complexity control problem, e.g. [9, 17–21]. One trivial method is to remove bars with cross-sectional areas below a predefined threshold value from the structure in a post-processing step [22]. The arbitrary choice of a threshold value yields an iterative procedure. Although this approach is very easy to implement, it bears high risks of leading to results where equilibrium is disrupted due to the removal of members.

Another approach is to introduce a penalty exponent $p > 1$ for the design variables, which aims at avoiding values unequal to 0 or 1 by reducing the stiffness assignment of intermediate values of the design variables [17, 18]. Although it is easy to implement, this measure usually destroys the convexity of the optimization problem. In order to limit the influence of the penalty exponent, it should therefore be increased continuously rather than starting already from its final value. Consequently, the final outcome is significantly dependent on the magnitude of p and the way it is increased. Moreover, the main drawback is that no direct control of the complexity is given.

A rather sophisticated approach is pursued by He and Gilbert [19] in a kind of simultaneous topology and shape optimization called "layout optimization", where additionally the node configuration is optimized in a post-processing step. Although the approach is mathematically convincing, it is numerically challenging and therefore not intuitive for practical applications. Furthermore, again, direct control of complexity is difficult.

If the objective function is to minimize the total weight, complexity control can be achieved by penalizing bar lengths, with shorter members contributing over-proportionally to the structure volume, see [20]. Although the linear character of the problem is maintained, the approach is not applicable to stiffness maximization problems with volume constraint.

In [9] and [21], a similar strategy by assigning costs to the struts is followed. The sum of all costs of active members is then added to the objective function. In doing so, the higher the cost value is chosen, the fewer bars are used by the optimization algorithm for the final structure. What is problematic when used in conjunction with minimizing the compliance as the objective function, however, is that consistency of units is not respected. As a result, the compliance term and the cost term can be of different magnitude, which can cause numerical problems. In turn, any kind of scaling of the single terms will influence the value of the bar costs to be chosen for achieving a result with a certain complexity. The magnitude of costs to be chosen is therefore always problem-specific.

An intuitive approach is to replace the volume constraint with a constraint on the total number of bars of which the optimal structure consists [9]:

$$g = \sum_{e=1}^{N_e} H(x_e) \leq N_e^* \tag{5.9}$$

In Eq. (5.9), $H(x_e)$ is a smooth approximation of the heaviside function, which takes values between 0 and 1 in order to count all "active" bars and N_e^* is the permissible total number of bars. The smooth approximation of the heaviside function reads

$$H(x_e) = \left[1 - \exp(-\beta_H x_e) + x_e \exp(-\beta_H)\right] \tag{5.10}$$

where β_H is a shape parameter. For $\beta_H = 0$, a linear interpolation is obtained, in which the value of a design variable contributes proportionally to the amount of active bars. With higher values for β_H, $H(x_e)$ approaches the heaviside step function ($\beta_H = \infty$), which counts bars with $0 > x_e \geq 1$ as "active" ($H(x_e) = 1$) and all others as "inactive" ($H(x_e) = 0$) as shown in Figure 5.7a. However, too large values for the shape parameter result in very steep gradients for low values of the design variables (Figure 5.7b), i.e. the step function is approximated too sharply, which can lead to numerical problems within the optimization. Consequently, choosing a reasonable value for the shape parameter is a compromise between numerical stability and sufficient differentiation between "active" and "inactive" bars. In this regard, $\beta_H = 100$ appears to be a suitable choice in most cases.

Besides the high degree of intuitiveness, a further advantage of this approach is that it favors a 0-1 distribution of the design variables, i.e. there are predominantly bars with $A_e(x_e) = A_{max}$ in the final structure. This is advantageous for developing STMs, because the required cross-sectional areas of rebars and compression struts are subsequently dimensioned based on the corresponding axial strut force and the material's utilizable stress, i.e. the yield stress of the reinforcement and the effective compressive strength of the concrete, respectively, and should not be determined by

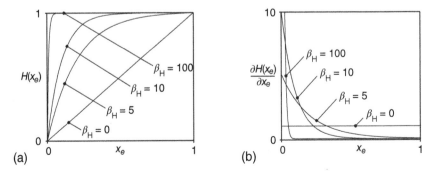

(a)

(b)

Figure 5.7 (a) Approximation of the heaviside step function for different shape parameters β_H, (b) related derivation functions.

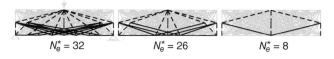

$N_e^* = 32$ $N_e^* = 26$ $N_e^* = 8$

Figure 5.8 Controlling the complexity of optimization results by limiting the number of allowed bars.

the optimization algorithm only with regard to stiffness maximization. For the sake of completeness, it should be noted that alternatively the number of nodes can also be limited instead of the number of members, as showed in [21].

The optimization problem with the constraint on the number of bars reads

$$\text{find: } \mathbf{x} = [x_1, x_2, \dots, x_{N_e}]^\mathsf{T}$$

$$\text{such that: } f = c(\mathbf{x})/c^0 = \sum_{e=1}^{N_e} \mathbf{u}_e^\mathsf{T} \mathbf{k}_e(x_e) \mathbf{u}_e/c^0 \to \min_{\mathbf{x}}$$

$$\text{subject to: } g = \sum_{e=1}^{N_e} H(x_e) - N_e^* \le 0 \tag{5.11}$$

$$\mathbf{K}(\mathbf{x})\mathbf{U} = \mathbf{F}$$

$$0 \le x_e \le 1 \qquad\qquad e \in [1, N_e]$$

where $\mathbf{u}_e$ and $\mathbf{k}_e$ are the element's displacement vector and stiffness matrix, respectively, $\mathbf{U}$ and $\mathbf{K}$ are the assembled global counterparts, and $\mathbf{F}$ is the global load vector. For numerical reasons, the objective function in Eq. (5.11) is related to the compliance value c^0 of the initial ground structure at iteration step $k = 0$. The complexity of the optimization result can be easily steered by adjusting the maximum allowable number of bars (N_e^*), as the example shown in Figure 5.8 demonstrates. Note that the result for $N_e^* = 8$ appears as if it consists of seven members, but the vertical strut at midspan is actually composed of two short bars arranged one above the other.

5.3.2 Sensitivity Analysis and Solving

In order to solve the optimization problem posed in Eq. (5.8), a gradient-based algorithm, e.g. the Method of Moving Asymptotes (MMA) [23], or an update scheme derived from an optimality criterion (OC) [12] can be used. Despite its heuristic character, the latter will be applied, as it operates very fast, is robust and reliable due to its tailored approach to the problem. It will be briefly introduced below.

First, however, the sensitivities of the objective and the constraint function are required, in other words, the functions' partial derivatives with respect to the design variables. For the minimum compliance problem, the sensitivities of the objective function can best be computed using the *adjoint method*. For a comprehensive derivation, the reader is referred to the related references, for instance [12, 24, 25] with detailed information. The sensitivity of the objective function reads

$$\frac{\partial f}{\partial x_e} = \left(\frac{\partial \mathbf{u}_e^\mathsf{T}}{\partial x_e} \mathbf{k}_e \mathbf{u} + \mathbf{u}_e^\mathsf{T} \frac{\partial \mathbf{k}_e}{\partial x_e} \mathbf{u}_e + \mathbf{u}_e^\mathsf{T} \mathbf{k}_e \frac{\partial \mathbf{u}_e}{\partial x_e} \right)/c^0 = \left(-\mathbf{u}_e^\mathsf{T} \frac{\partial \mathbf{k}_e}{\partial x_e} \mathbf{u}_e \right)/c^0 \tag{5.12}$$

where the derivative of the local stiffness matrix is computed straightforwardly as:

$$\frac{\partial \mathbf{k}_e}{\partial x_e} = \left(1 - x_{\min}\right) A_{\max} L_e \mathbf{k}_e^0 \tag{5.13}$$

To enhance readability, the notation of dependencies in Eqs. (5.12) and (5.13) is omitted. The sensitivity of the constraint function can be computed directly and follows to:

$$\frac{\partial g}{\partial x_e} = \beta_H \exp(-\beta_H x_e) + \exp(-\beta_H) \tag{5.14}$$

Again, a simplified notation is used in Eq. (5.14).

In order to derive an OC, the Lagrangian function is employed:

$$L = \mathbf{U}^\top(\mathbf{x})\mathbf{K}(\mathbf{x})\mathbf{U}(\mathbf{x}) + \Lambda\left(\sum_{e=1}^{N_e} H(x_e) - N_e^*\right) + \sum_{e=1}^{N_e} \lambda_e\left(0 - x_e\right) + \sum_{e=1}^{N_e} \gamma_e\left(x_e - 1\right) \tag{5.15}$$

where $\Lambda \geq 0$, $\lambda_e \geq 0$, and $\gamma_e \geq 0$ are the Lagrangian multipliers for $e \in [1, N_e]$. The stationarity condition of the Lagrangian function leads to

$$\frac{\partial L}{\partial x_e} = \left(\frac{\partial \mathbf{U}^\top}{\partial x_e}\mathbf{K}\mathbf{U} + \mathbf{U}^\top\frac{\partial \mathbf{K}}{\partial x_e}\mathbf{U} + \mathbf{U}^\top\mathbf{K}\frac{\partial \mathbf{U}}{\partial x_e}\right) + \Lambda\frac{\partial g}{\partial x_e} - \lambda_e + \gamma_e \overset{!}{=} 0 \tag{5.16}$$

Using the general FE equation $\mathbf{K}\mathbf{U} = \mathbf{F}$, it can be shown that

$$\mathbf{U}^\top\mathbf{K}\frac{\partial \mathbf{U}}{\partial x_e} = -\mathbf{U}^\top\frac{\partial \mathbf{K}}{\partial x_e}\mathbf{U} \tag{5.17}$$

Substituting Eq. (5.17) into the stationarity condition leads to

$$\frac{\partial L}{\partial x_e} = \underbrace{-\mathbf{U}^\top\frac{\partial \mathbf{K}}{\partial x_e}\mathbf{U} + \Lambda\frac{\partial g}{\partial x_e}}_{\dfrac{\partial f_0}{\partial x_e}} - \lambda_e + \gamma_e \overset{!}{=} 0 \tag{5.18}$$

Finally, after reshaping Eq. (5.18), the OC is obtained:

$$\underbrace{\frac{-\dfrac{\partial f}{\partial x_e}}{\Lambda\dfrac{\partial g}{\partial x_e}}}_{G_e} = 1 - \lambda_e + \gamma_e \tag{5.19}$$

By evaluating the above equation, an update scheme for the design variables can be derived, see [12, 26, 27]:

$$x_e^{(k+1)} = \begin{cases} \left[G_e^{(k)}\right]^{0.5} x_e^{(k)} & \text{if } M_x^- \leq \left[G_e^{(k)}\right]^{0.5} x_e^{(k)} \leq M_x^+ \\ M_x^- & \text{if } M_x^- \geq \left[G_e^{(k)}\right]^{0.5} x_e^{(k)} \leq M_x^+ \\ M_x^+ & \text{if } M_x^- \leq \left[G_e^{(k)}\right]^{0.5} x_e^{(k)} \geq M_x^+ \end{cases} \tag{5.20}$$

where an exponent of 0.5 is introduced for numerical damping. For the lower and upper bounds it holds that

$$M_x^- = \max\left\{(1 - \mu)x_e^{(k)}, x_e^L\right\} \tag{5.21a}$$

$$M_x^+ = \min\left\{x_e^U, (1 + \mu)x_e^{(k)}\right\} \tag{5.21b}$$

In Eq. (5.21), $\mu = 0.2$ is a move limit to prevent too large changes of the design variables between two iterations, which may cause convergence problems, whereas $x_e^L = 0$ and $x_e^U = 1$ are the lower and upper bounds for the design variables. The Lagrangian multiplier Λ is determined numerically, for example by using the bisection method. This is done in such a way that the constraint is attained in the optimum, since it can be assumed that the stiffest structure (= lowest compliance) is formed by utilizing the largest permissible number of members.

5.3.3 Optimization Process

The procedure of TTO is shown in the flow chart in Figure 5.9. First, the FE model (analysis model), i.e. the design space, the acting loads and the restraints, as well as the optimization problem, are initialized. A FE analysis is then conducted with the aim of computing all node displacements. Using the latter, the values of the objective and constraint function and their sensitivities with respect to the design variables are calculated. The sensitivities are employed for updating the design variables with the update scheme presented in Section 5.3.1 based on the OC derived from the Lagrangian function. This completes the current iteration step and initiates the subsequent one by increasing the iteration number by 1.

The convergence criterion is now being checked. For example, a possible stop criterion is the largest difference between the design variables of two subsequent iterations:

$$\max \Delta x_e = \max \left| x_e^{(k+1)} - x_e^{(k)} \right| \leq \text{tol}_x \tag{5.22}$$

for all $e \in [1, N_e]$, where the threshold value could be set, for example, to $\text{tol}_x = 10^{-5}$. To avoid oscillating infinite loops or excessive computing times, it is reasonable to introduce a maximum number of iterations as an additional criterion and thereby limit the optimization to, for instance, $k_{\max} = 300$ iterations [17].

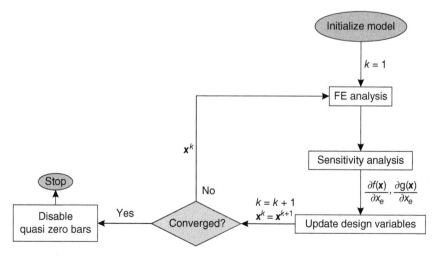

Figure 5.9 Flowchart of the TTO approach with complexity control.

If neither of the two criteria is met, the next calculation run starts by performing a new FE analysis based on the updated design variables. Otherwise, the optimization is stopped and all quasi-zero struts, meaning members whose design variables are below a predefined threshold value, are disabled for visualization in a post-processing step. For the examples given in Section 5.5, the threshold value, determining whether a bar is displayed or not, is defined as follows:

$$\alpha_{min}^{vis} = 0.005 > \frac{A_e}{\max{(\mathbf{A})}} \tag{5.23}$$

where α_{min}^{vis} is the threshold value for visualization, A_e is the cross-sectional area of member e, and $\mathbf{A}$ is a vector containing the cross-sectional areas of all struts. The maximum value in $\mathbf{A}$ is taken for comparison. In doing so, all members with an area A_e which is less than 0.5 % of the maximum member's cross-sectional area of all elements will not be visualized.

5.3.4 Recommendations for Practical Application

For the practical application of the method, this section provides instructions and recommendations for a use. They are given consecutively, starting with the general recommendations on how to set up the optimization problem most reasonably.

5.3.4.1 Setting Up the Optimization Problem

Model Definition The way in which the boundary conditions are defined, i.e. loads and constraints, the base mesh, the level of connectivity (Lvl), and consequently the ground structure strongly influence the result. For this reason, the user must carefully consider suitable modeling in advance. This is an engineering task, may require some experience and is comparable, for example, to setting up a FE model.

System modeling should aim at rather coarse base meshes. However, it should be noted that secondary local effects could be disregarded if the discretization is too coarse. The guiding principle for the base mesh should therefore be "as coarse as possible, as fine as necessary". If the resulting STMs of a coarse base mesh is preferred over a finer one, potential secondary effects, which are apparent in the fine but not in the coarse one, must in any case be considered structurally in the design. Moreover, generally, a uniform base mesh with equally spaced nodes should be aimed for. On the other hand, in accordance with the aforementioned basic modeling principle, the node density should be adaptive to stress gradients, so specifically denser at critical regions and more coarse at noncritical ones. In this way, local effects can be taken into account while at the same time unnecessarily complicated shaping of the STMs is counteracted. In this procedure, parallels can be drawn with standard FE modeling and recognized mesh adaption methods like h-, p-, or hp approaches [16, 28, 29]. Loads and restraints should be modeled in a simple manner and be aligned with the guidelines for the development of STMs according to [2]. The most crucial ones are

- merge distributed loads into resultant forces, which, for example, act at the quarter points of the load distribution range (Figure 5.10a)
- model merely the discontinuity region to be examined and take into account the resulting compressive and tensile forces (=reinforcement locations) of adjacent structural regions as loads or supports (Figure 5.10b)

- avoid unnecessary restraints, because the computed STMs strives to interconnect all load acting points and fixed boundaries (Figure 5.10c)

Level of Connectivity (Lvl) On the one hand, the maximum possible solution space of a base mesh results when using the highest possible value for *Lvl*. On the other hand, the associated STMs are often complicated and unsuitable for practical use, because they may contain overlapping crossing struts or struts having incompatible inclinations.

This is partly due to the way the struts of a ground structure are defined: to avoid collinear overlapping struts, shorter members are preferred over longer ones as described in [22], see Figure 5.11a. Partly, of course, this is also due to the inherent tendency toward more complex structures with increasing .values on *Lvl*. As a

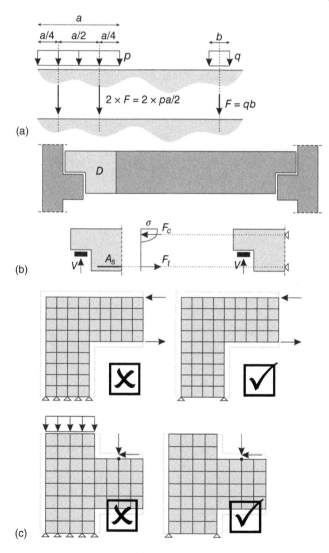

Figure 5.10 Recommendations for modeling STMs optimization problems.

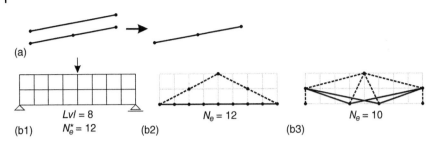

(a)

(b1) $Lvl = 8$ (b2) $N_e = 12$ (b3) $N_e = 10$
        $N_e^* = 12$

Figure 5.11 (a) Avoiding collinearity of overlapping bars at $Lvl > 1$, (b1) optimization problem, (b2) expected STMs, (b3) optimization result with inclined bars.

result, inclined long struts are preferred to horizontal or vertical "strut chains," as the latter require higher numbers of struts and with the former; therefore, higher bracing (and thus stiffness) is possible though formally using the same or even a lower amount of members (Figure 5.11b). Hence, it is advisable to compute results for both maximum and minimum ($Lvl = 1$) – or even for intermediate – values of levels of connectivity and then choose a suitable STMs.

Visualization For visualizing the final result, members whose design variables are low (quasi-zero struts) should be excluded and the widths of the remaining "active" members should be scaled according to their associated design variables. For example, all members with a cross section $A_e(x_e) < \alpha_{min}^{vis} \max(\mathbf{A})$ for all $e \in [1, N_e]$ are hidden in all results presented, where $\max(\mathbf{A})$ is the largest cross section available. Typically, it holds that $\max(\mathbf{A}) = A_{max}$. The threshold value used here corresponds to $\alpha_{min}^{vis} = 0.005$ (see Eq. (5.23)).

Choice of A_{max} A_{max} hardly influences the qualitative result and has no relevance for the STMs, since, on the one hand, the presented approach favors a 0-1 distribution of the design variables and, on the other hand, the stress limits are not taken into account within the optimization. The respective required cross-sectional areas of the struts and ties must be determined at a later stage according to the design principles given in [2]. Hence, A_{max} should likely be chosen within the limits that are physically reasonable.

Initial Design Variables Distribution Since the problem is non-convex, it should be noted that the initial choice of the design variables can have an influence on the result. However, for the same reasons, it is hardly possible to define good starting points to reach the global optimum. For all results presented here, a homogeneous initial distribution of $x^{(0)} = 0.1$ was applied.

Number of Iterations To avoid premature termination of the optimization, a minimum number of iterations should be defined, for example $k_{min} = 30$. Conversely, for practical reasons, the duration of computations should be limited, for example to $k_{max} = 300$ iterations. Both limits were adopted when calculating the examples shown herein.

Shape Parameter β_H The shape parameter β_H determines how closely the heaviside step function is approximated. Larger values yield a better differentiation between "active" and "inactive" bars, but lead to steep gradients for small values of the design variables, which can cause numerical problems. A sound compromise between sufficient approximation and numerical stability may be, for instance, $\beta_H = 100$. Depending on the case, the shape parameter can be adjusted by experienced users as required.

5.3.4.2 Procedure

Figure 5.12 illustrates the recommended procedure for the optimization-based determination of STMs using TTO. After modeling the design space along with all

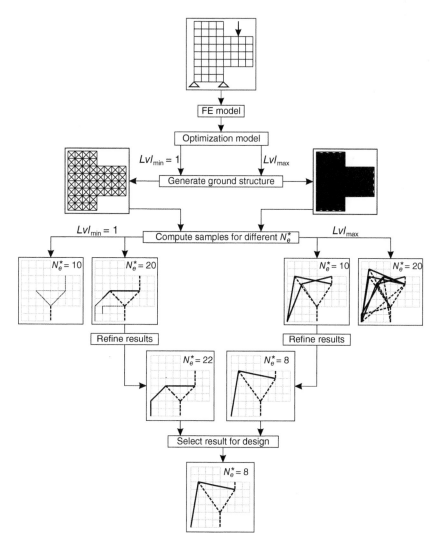

Figure 5.12   Recommended procedure for the STMs striving TTO approach.

loads and restraints, the ground structure is generated using $Lvl = Lvl_{max}$. Samples are now computed for different limits of the constraint function, for instance, $N_e^* = 10, 20, 30$. These results serve as "sampling points" and are used to determine a reasonable value range for the constraint. Starting from the most promising result, further structures are computed by adjusting the permitted number of bars to further resolve (increase N_e^*) or simplify (decrease N_e^*) the STMs. Following the same procedure, STMs can be calculated for $Lvl = 1$ as well as for additional, intermediate levels of connectivity, if needed. Lastly, the user compares the STMs resulting from the different values for Lvl and decides on the most adequate one for application. The axial forces of the members can be directly read off as bar forces. They are then used in a post-processing step for the design.

5.4 Continuum-Truss Topology Optimization (CTTO) Approach

In certain applications, such as partial area loading, neither TTO (Figure 5.13b) nor CTO (c) provide valid results resembling the classical STMs (a). In case (b) only longitudinal struts arise neglecting the lateral splitting. Similar to that, in case (c), no tensile struts form. Indeed, topology optimization approaches are not sufficiently capable of representing lateral tensile stress induced by load spreading and only concentrate on the prevailing compressive stress. In doing so, the pronounced anisotropic behavior of concrete, its comparatively low tensile strength, and the interaction with the embedded steel reinforcement is not taken into account. For problems where it is necessary to include these factors, hybrid continuum-truss topology optimization (CTTO) approaches can provide suitable results, cf. Figure 5.13d.

CTTO implementations were first proposed by Amir and Bogomolny [18, 30, 31]. They include a damage model, where concrete and steel are modeled as elastoplastic

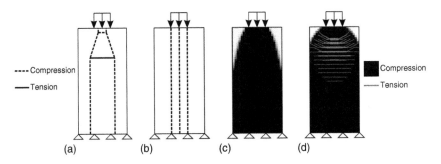

Figure 5.13 Partial area loading: (a) classical STMs, (b) TTO, (c) CTO, (d) combined continuum-truss approach, adapted from [17].

materials using the Drucker–Prager yield criterion [32]. About the same time, the so-called "embedded truss formulation" was developed by Gaynor et al., which is a more simplified approach since only bilinear material models are applied [33–35]. Smarslik and Mark further enhanced the embedded truss formulation focusing on partial area loading in tunnel lining segments with an efficient post-processing approach to determine rebar layouts and to consider steel fiber reinforcement as well as a combination of both [17, 36–38]. This method will be further discussed hereafter.

5.4.1 Problem Statement

Two finite element types are introduced simultaneously within the optimization problem, namely truss and continuum elements. The element types represent the different materials, that is the trusses to correspond to the steel rebars and the continua to the concrete volume. They also define the two separate but superimposed design spaces, namely Ω_s containing the truss elements and Ω_c consisting of the continuum elements (Figure 5.14a). Each design space also contains one separate set of design variables each: **a** for Ω_s and ρ for Ω_c. For the former, the design variables correspond directly to the cross-sectional areas, hence it holds $A_e = a_e$. For the latter, they correspond to the elements relative densities and thereby control the material assignment, c.f. Eq. (4.3), of each.

The nodes of Ω_s and Ω_c coincide, thus interconnecting the design spaces, which allows transmission of forces between them. For the sake of simplicity, a nonslip condition is assumed for the rebars. The mesh of Ω_c defines the number as well as the location of the nodes. To describe the way the design spaces are superimposed, the so-called block size (*BS*) is introduced. *BS* is a parameter that determines how many continuum elements are connected to form a square block whose corners correspond to the nodes of the base mesh for Ω_s (Figure 5.14b). *BS* takes integer values, where *BS* > 1 implies that the ground structure is not connected to each node of Ω_c. Given a constant level of connectivity, the number of bars forming the ground structure decreases with increasing *BS*.

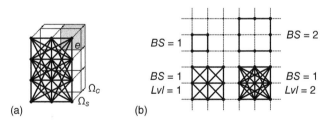

(a) (b)

Figure 5.14 CCTO: (a) superimposed design domains, (b) ground structures depending on the Level of Connectivity (*Lvl*) and block size (*BS*), adapted from [17].

The optimization scheme is set up as a compliance minimization problem with volume constraint by combining both design spaces:

$$\text{find:} \quad \mathbf{a} = \left[a_1, a_2, \ldots, a_{N_s}\right]^{\mathsf{T}}, \rho = \left[\rho_1, \rho_2, \ldots, \rho_{N_c}\right]^{\mathsf{T}}$$

$$\text{such that:} \quad f = c(\mathbf{a}, \rho) = \mathbf{U}^{\mathsf{T}} \mathbf{K}(\mathbf{a}, \rho, \sigma_s, \sigma_c)\mathbf{U}$$

$$= \underbrace{\sum_{e=1}^{N_s} \mathbf{u}_e^{\mathsf{T}} \mathbf{k}_{e,s}(a_e, \sigma_{e,s})\mathbf{u}_e}_{c_{e,s}} + \underbrace{\sum_{e=1}^{N_c} \mathbf{u}_e^{\mathsf{T}} \mathbf{k}_{e,c}(\rho_e, \sigma_{e,c})\mathbf{u}_e}_{c_{e,c}} \to \min_{\mathbf{a},\rho}$$

$$\text{subject to:} \quad g = \sum_{e=1}^{N_s} a_e L_e + \sum_{e=1}^{N_c} \rho_e v_e = V \le \beta V^0$$

$$\mathbf{K}(\mathbf{a}, \rho, \sigma_s, \sigma_c)\mathbf{U} = \mathbf{F}$$

$$0 < a_{\min} \le a_e \le a_{\max} \qquad\qquad\qquad e \in [1, N_s]$$

$$0 \le \rho_e \le 1 \qquad\qquad\qquad\qquad\quad e \in [1, N_c]$$

$$(5.24)$$

In Eq. (5.24), $\mathbf{a}$ is a vector containing the truss design variables a_e of the design space Ω_s. $a_{\max}$ acts as their upper bound, which is to be chosen based on reasonable reinforcing. $a_{\min} = 10^{-6}$ cm^2 is a nonzero lower bound to avoid numerical singularity. 10^{-6} cm^2 undergoes usual reinforcement bar areas by at least five magnitudes and thus can easily be neglected. The design variables ρ_e in Ω_c are stored in the vector ρ. $\mathbf{F}$ is the global load vector, and $\mathbf{U}$ and $\mathbf{u}_e$ are the global and local displacement vector, respectively. $\mathbf{k}_{e,s}$ is the element stiffness matrix of steel and $\mathbf{k}_{e,c}$ is the element stiffness matrix of concrete, whereas $\mathbf{K}$ is the assembled global stiffness matrix. N_s is the number of truss elements in the ground structure, N_c is the total number of continuum elements, L_e is the truss length, v_e is the element volume, V is the total volume, V^0 is the total initial volume composed of all trusses and continua, $\beta \in [0,1]$ is the ratio of actual to initial volume, σ_s is a vector containing all truss forces $\sigma_{e,s}$, and σ_c is a matrix containing the principal stresses $\sigma_{e,c}$ of all continuum elements. The way the stiffness matrices depend on the element stresses is described next.

The materials are represented by bilinear constitutive models, which are defined by a Young's modulus and a Poisson's ratio in compression and tension as shown in Figure 5.15a. Here, steel is assumed as tension-*only* material with $E_s^+ = 210\,000$ MPa under tension and $E_s^- = 0$ MPa under compression. It should be noted that, of course, steel can overtake compressive stresses, but this bearing reaction should intentionally be suppressed to steer all compressive stress components into the concrete continuum. In contrast, concrete is assumed as compression-*dominant* material and exhibits a compressive Young's modulus of $E_c^- = 30\,000$ MPa. The tensile Young's modulus is set considerably lower at, for instance, $E_c^+ = E_c^-/100 = 300$ MPa. This allows it to carry minor tensile stresses, which ensures non-singularity of the global stiffness matrix and at the same time does not affect the results in a noticeable extent.

Figure 5.15 (a) Bilinear stress–strain relationship for concrete and steel, (b) stress-dependent material parameters for continuum elements, adapted from [17].

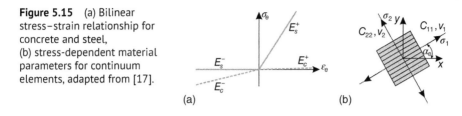

(a)   (b)

In the first iteration, a linear elastic, isotropic FE analysis is performed from which the displacements, truss forces, and stresses are computed. The information is used to determine the initial material properties, namely the stress-dependent Young's modulus, of each truss and continuum element, whereupon the element stiffness matrices are modified accordingly. In the case of steel,

$$\mathbf{k}_{e,s}(a_e, \sigma_{e,s}) = a_e^{p_s} \frac{E_s(\sigma_{e,s})}{L_e} \mathbf{R}_e \tag{5.25}$$

applies, where $\sigma_{e,s}$ is the truss' axial stress and p_s is an optional penalization factor. The latter can be adjusted moderately between 1.0 and 1.2 in order to reduce complexity of the bar layout [17]. $\mathbf{R}_e$ is the rotational matrix

$$\mathbf{R}_e = \begin{bmatrix} l^2 & lm & -l^2 & -lm \\ lm & m^2 & -lm & -m^2 \\ -l^2 & -lm & l^2 & lm \\ -lm & -m^2 & lm & m^2 \end{bmatrix} \tag{5.26}$$

with $l = \cos(\alpha)$ and $m = \sin(\alpha)$, where α is the angle between the bar and the horizontal axis (Figure 5.5). The Young's modulus is chosen depending on the sign of the truss stress:

$$E_s(\sigma_{e,s}) = \begin{cases} E_s^+ & \text{for } \sigma_{e,s} > 0 \\ E_s^- & \text{for } \sigma_{e,s} \leq 0 \end{cases} \tag{5.27}$$

For concrete, anisotropy is introduced by the orthotropic model of Darwin and Pecknold [39, 40], whereas, however, only the bilinear elastic portion is employed. In doing so, the element stiffness depends on the principal stresses along with their orientation, cf. Figure 5.15b. This means full stiffness in the direction of compressive stresses and a reduced stiffness in the direction of tensile stresses. As a consequence, tensile stresses are transferred to the truss ground structure. The element stiffness matrix reads

$$\begin{aligned} \mathbf{k}_{e,c}(\rho_e, \sigma_{e,c}) &= (\rho_e^{p_c} + \rho_{\min})\mathbf{k}_{e,c,0}(\mathbf{C}_e(\sigma_{e,c})) \\ &= (\rho_e^{p_c} + \rho_{\min})\int_{\Omega_e} \mathbf{B}_e^\top \mathbf{C}_e(\sigma_{e,c})\mathbf{B}_e dV \end{aligned} \tag{5.28}$$

where $\mathbf{C}_e(\sigma_{e,c})$ is the stress-dependent constitutive matrix, $\mathbf{B}_e^\top$ is the B-operator containing the partial derivatives of the shape function with respect to the global coordinates for all nodes of the element, and p_c is a penalty exponent which is usually set to $p_c = 3$ to favor a 0-1 distribution of the design variables, see Section 4.1.1. The constitutive matrix reads

$$\mathbf{C}_{e,\alpha} = \begin{bmatrix} C_{11} & v_{\text{eff}}C_{11}C_{22} & 0 \\ v_{\text{eff}}C_{11}C_{22} & C_{22} & 0 \\ 0 & 0 & \frac{1}{4}\left(C_{11} + C_{22} - 2v_{\text{eff}}\sqrt{C_{11}C_{22}}\right) \end{bmatrix} \quad (5.29)$$

with the index α indicating that $\mathbf{C}_{e,\alpha}$ is defined locally on the principal plane. C_{ii} is the directional Young's modulus depending on the principal stresses $\sigma_{e,c,i}$ where i stands for 1,2 labeling the two principal directions:

$$C_{ii} = \begin{cases} E_c^+ & \text{for } \sigma_{e,c,i} \geq 0 \\ E_c^- & \text{for } \sigma_{e,c,i} < 0 \end{cases} \quad (5.30)$$

For the effective Poisson's ratio, it holds that

$$v_{\text{eff}} = \sqrt{v_1 v_2} \quad (5.31)$$

with

$$v_i = \begin{cases} v_c^+ = v_c^- \frac{E_c^+}{E_c^-} & \text{for } \sigma_{e,c,i} \geq 0 \\ v_c^- & \text{for } \sigma_{e,c,i} < 0 \end{cases} \quad (5.32)$$

where v_c^+ and v_c^- are the Poisson's ratios under tension and compression, respectively. $\mathbf{C}_{e,\alpha}$ can then be transposed into global coordinates for the assembly of the global stiffness matrix:

$$\mathbf{C}_e = \mathbf{Q}_e^T \mathbf{C}_{e,\alpha} \mathbf{Q}_e \quad (5.33)$$

with

$$\mathbf{Q}_e = \begin{bmatrix} l^2 & m^2 & lm \\ m^2 & l^2 & -lm \\ -2lm & 2lm & l^2 - m^2 \end{bmatrix} \quad (5.34)$$

where $l = \cos(\alpha)$ and $m = \sin(\alpha)$ (Figure 5.15b). Finally, the global stiffness matrix is assembled from the local stiffness matrices of all truss and continuum elements. In doing so, the two domains Ω_s and Ω_c are interlinked.

5.4.2 Sensitivity Analysis and Solving

The sensitivities can be computed following the adjoint method [12, 25]. Regarding the truss elements, this gives

$$\frac{\partial f}{\partial a_e} = -p_s a_e^{(p_s-1)} \mathbf{u}_e^T \frac{E_s(\sigma_{e,s})}{L_e} \mathbf{R}_e \mathbf{u}_e \quad (5.35\text{a})$$

$$\frac{\partial g}{\partial a_e} = \frac{L_e}{\beta V^0} \quad (5.35\text{b})$$

For the continuum elements, it leads to

$$\frac{\partial f}{\partial \rho_e} = -p_c \rho_e^{(p_c-1)} \mathbf{u}_e^\mathsf{T} \mathbf{k}_{e,c,0} \left(\mathbf{C}_e \left(\sigma_{e,c} \right) \right) \mathbf{u}_e \tag{5.36a}$$

$$\frac{\partial g}{\partial \rho_e} = \frac{v_e}{\beta V^0} \cdot \tag{5.36b}$$

Since the bilinear material models require an iterative procedure to find equilibrium, two loops are to be executed: an outer optimization loop to update the design variables and an inner FE loop to solve the equilibrium equations (cf. Figure 5.18). For the latter, a set of convergence criteria can be formulated regarding the recommendations given in [17, 33]. Accordingly, equilibrium is obtained if

- < 0.1 % of all trusses change between tension and compression AND
- the principal plane orientation change is < 0.01°.

Experience shows that, beyond the above stated criteria, it is reasonable to limit the maximum number of iterations to avoid oscillating behavior or very slow convergence without significant improvement. 10 iterations make a good choice for this purpose.

Once equilibrium is achieved, the sensitivities can be calculated to subsequently update the design variables α and ρ. A filter method (Section 4.1.3) must be applied to the continuum elements in order to circumvent the well-known numerical problems [12, 41, 42]. The design variables can be updated with a nonlinear optimization algorithm such as MMA. Alternatively, a well-suited OC-based heuristic update scheme can be derived, as is the case for CTO and TTO. For a more detailed discussion, see [17]. Accordingly, the stationarity condition of the Lagrangian function leads to the well-known optimality conditions:

$$G_{e,s} = -\frac{\frac{\partial f}{\partial a_e}}{\Lambda \frac{\partial g}{\partial a_e}} = 1 - \lambda_e + \gamma_e \tag{5.37}$$

and

$$G_{e,c} = -\frac{\frac{\partial f}{\partial \rho_e}}{\Lambda \frac{\partial g}{\partial \rho_e}} = 1 - \lambda_e + \gamma_e \tag{5.38}$$

for the truss and the continuum elements, respectively, where $\lambda_e \geq 0$, $\gamma_e \geq 0$, and $\Lambda \geq 0$ are the Lagrange multipliers. Following the explanations in [12], heuristic update schemes can be developed by evaluating the optimality conditions with respect to the design variables. For the truss elements, this leads to:

$$a_e^{(k+1)} = \begin{cases} M_a^- & \text{if} & \left[G_{e,s}^{(k)} \right]^{0.5} a_e^{(k)} \leq M_a^- \\ M_a^+ & \text{if} & \left[G_{e,s}^{(k)} \right]^{0.5} a_e^{(k)} \geq M_a^+ \\ \left[G_{e,s}^{(k)} \right]^{0.5} a_e^{(k)} & \text{else} \end{cases} \tag{5.39}$$

where

$$M_a^- = \max\left\{(1 - \mu_a)\, a_e^{(k)}, a_{\min}\right\} \tag{5.40a}$$

$$M_a^+ = \min\left\{a_{\max}, (1 + \mu_a)\, a_e^{(k)}\right\} \tag{5.40b}$$

Similarly, for the continuum elements, the update scheme reads

$$\rho_e^{(k+1)} = \begin{cases} M_\rho^- & \text{if} & \left[G_{e,c}^{(k)}\right]^{0.5} \rho_e^{(k)} \le M_\rho^- \\ M_\rho^+ & \text{if} & \left[G_{e,c}^{(k)}\right]^{0.5} \rho_e^{(k)} \ge M_\rho^+ \\ \left[G_{e,c}^{(k)}\right]^{0.5} \rho_e^{(k)} & \text{else} \end{cases} \tag{5.41}$$

with

$$M_\rho^- = \max\left\{(1 - \mu_\rho)\, \rho_e^{(k)}, 0\right\} \tag{5.42a}$$

$$M_\rho^+ = \min\left\{1, (1 + \mu_\rho)\, \rho_e^{(k)}\right\} \tag{5.42b}$$

Here, 0.5 is a numerical damping exponent, k is the current iteration number, whereas μ_a and μ_ρ are the move limits. Even though they may be chosen differently, 0.2 is a reasonable choice for both. Λ is equal for both optimality conditions and can be obtained using a bisection algorithm in such a way that the volume constraint limit is met. The design variables a and ρ are based on the same displacement field but are updated independently from each other. In [17], numerical studies reveal that the OC-based update schemes converge much faster and are more robust than general optimizers such as MMA, its globally convergent version globally convergent method of moving asymptotes (GCMMA) as well as an inner point method, all applied with default values.

CTTO is a complex approach and therefore rather challenging in its application. It demands in-depth knowledge from the user. Its performance depends on the choice of the initial parameters, which is more sensitive to changes than CTO and TTO. Hence, they have to be adapted problem-specifically. For a more detailed discussion and some general recommendations, see [17]. For this reason, it is advisable to apply CTTO only in exceptional cases, for example, if neither TTO nor CTO lead to viable results or for sophisticated applications, such as the development of hybrid reinforcement layouts containing bar and fiber reinforcements in partial area loading. The latter will be discussed as application example in Section 5.6.2.

5.4.3 Post-Processing

The design variables a of the truss elements are merely virtual cross-sectional areas which are only relevant within the optimization process, since no information regarding the material yield stress is considered. In order to assign a physical meaning, stress or geometry constraints should be taken into account within the optimization process, which would significantly increase complexity. Smarslik [17], therefore, proposed an integrated post-processing algorithm that allows a quantitative evaluation of the optimization results.

The virtual cross-sectional areas can be converted into physical ones by using a scaling factor which is computed based on the stiffness ratio between the bars and the continuum elements. The validity of this simple method can be proven by FE calculations, see [17]. Once the physical cross-sectional areas of the bars are obtained, the next step is to simplify the truss layout. Due to the bilinear material model used, some trusses exhibit only very low tensile forces, making them cost-inefficient in practice. In addition, the nonzero tensile Young's modulus assigned to the concrete avoids singular stiffness matrices but also makes the appearance of low tensile stresses in the continuum elements possible, which are not covered by the reinforcement. Both issues can be adequately addressed by taking into account the low tensile strength of concrete. Hence, low utilized bars can be removed by transferring their tensile forces to the adjacent continuum elements (Figure 5.16). In doing so, it also allows the incorporation of steel fiber reinforcement. For this purpose, the catchment area of each truss element within which its force is "smeared" into distributed stress must first be determined. The idea is to cover shares of distributed tensile stresses by the distributed tensile bearing ability of fibers that interconnect over cracks and thus ensure a certain tensile stress-bearing ability. The residual tensile stresses are steered to the trusses and thus lumped to single reinforcement bars.

The catchment areas can be defined as Voronoi areas [43, 44]. This allows all or some of a bar's tensile force to be transferred to the fiber effect in concrete to the extent the tensile strength provided steel fiber-reinforced concrete (FRC). This makes the procedure a valuable tool in cases that steel fibers are used to completely or partially replace conventional rebars.

Hence, two stress limits are introduced for continuum elements, cf. Figure 5.17a. The first is a very small share of the concrete's tensile strength (i.e. $\leq 0.1 f_{ctm}$) to eliminate poorly loaded trusses. The second is optional and represents the fiber reinforcement's uniaxial tensile design strength (f_{ctd}^f) and replaces the minimum value if fiber reinforcement is employed. Then, the fibers cover stresses until f_{ctd}^f. Those

Figure 5.16 Principle of truss layout simplification [17]: (a) initial result, (b) trusses of minor loading removed.

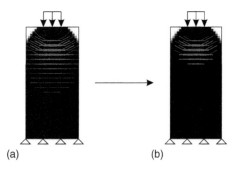

(a)   (b)

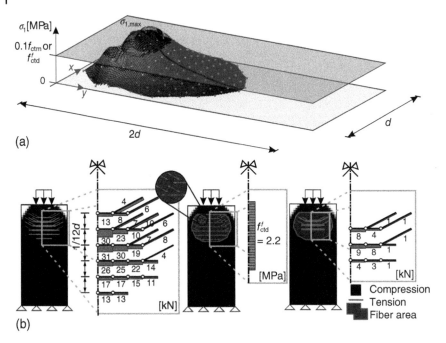

(a)

(b)

Figure 5.17 Post-processing the CTTO results: (a) stress limits in continuum elements for plain or fiber-reinforced concrete, (b) force transfer of the trusses into the concrete, adapted from [17].

exceeding f^f_{ctd} must be accounted for by conventional rebars. In this way, the reinforcement layout is simplified, the structural area relevant for FRC is identified, and the optimal fiber orientation is specified through the direction of the principal tensile stresses (Figure 5.17b).

5.4.4 Optimization Process

Figure 5.18 shows the flow chart of the CTTO approach. After initializing the models, the FE analysis is conducted. Since the continuum elements employ nonlinear material models, an inner loop has to be performed in order to solve the equations. Then, the sensitivities of the objective and the constraint functions are determined with respect to all design variables. Subsequently, the design variables are updated via the OC-based update scheme. Finally, it is determined whether the convergence criterion is met. If so, all almost zero bars are disabled and the optimization process is finished. For practical application, post-processing can then be performed, as described in the previous section. If convergence is not met, the next loop starts solving the FE equations based on the evolved design variable distribution from the last iteration step.

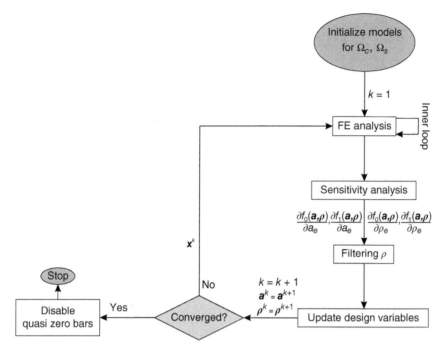

Figure 5.18 Flow chart of the CTTO approach for solutions.

5.5 Examples

5.5.1 CTO Approach

Example 5.1 (Deep beam 1). Figure 5.19a shows a simply supported square deep beam with dimensions of $L = H = 400$ cm which is loaded by a uniformly distributed load $q = 300$ kN/m. In the analysis model, q is represented by its two resultants $qL/2 = 600$ kN acting at the quarter points (b). The design space is discretized by $N_e = 80 \times 80 = 6400$ quadrilateral elements. The Young's modulus is set to $E = 30\,000$ MPa and the Poisson's ratio v equals 0.2.

Figure 5.19c shows the optimization results for different values of β using a filter radius of $r_{\min} = 1.5$ which refers to the size $a_e = 5$ cm of one finite element length. The results show that for $\beta = 0.2$ and $\beta = 0.4$ basic structures for STMs can be found, while higher values of the volume fraction yield material distributions that are rather unsuitable for identifying individual struts. On the basis of the result using 20 % of the initial material, the STMs in Figure 5.19d is geometrically derived by hand and then its member forces are computed. The STMs is very similar to the established model by Schlaich and Schäfer [45], cf. Figure 5.25a.

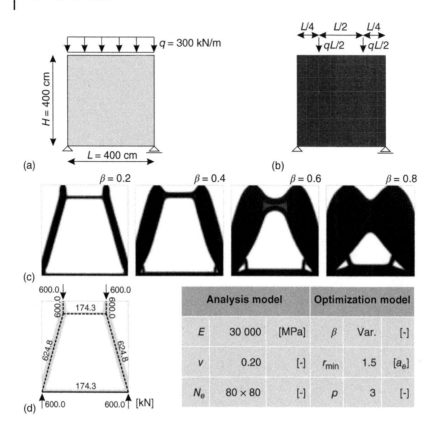

Figure 5.19 (a) Simple deep beam, (b) analysis model, (c) CTO results, (d) STMs.

Example 5.2 (Wall with block-outs). The two-span deep beam in Figure 5.20a is loaded by two point loads $F = 1$ MN at each midspan. With a total of four large openings that distort the flow of forces, it represents a typical practical case illustrating the difficulty of finding a compatible STMs for walls. The design space consists of a regular mesh of $N_e = 27\,000$ quadrilateral elements minus removed elements within the material free block-outs. The material parameters equal the values used in the previous example.

Figure 5.20b shows the calculated material distributions for three different volume fractions. In the result for $\beta = 0.1$, the single struts of the structure can be identified easily. However, the corresponding STMs (c) proves to be kinematically indifferent. The STMs is therefore manually supplemented by an additional top truss between the two loads in such a way that the kinematic determination is ensured. Information from the optimization results obtained by using higher values on β can be employed for this purpose, where an additional horizontal bracing appears at the upper edge. With this tie enhanced, the STMs becomes kinematically determined and its member forces can be computed. Figure 5.20d depicts the latter in relation to $F = 1$ MN. The complementary strut obviously takes only minor loads, which

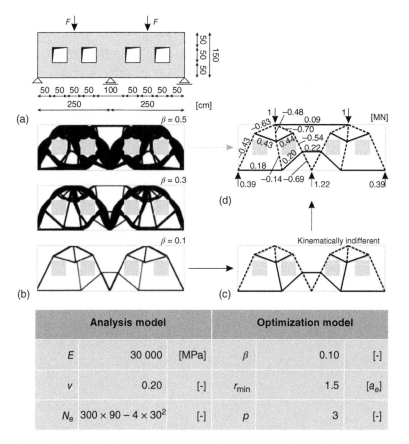

Figure 5.20 (a) Continuous deep beam with openings problem, (b) CTO results, (c) kinematically indeterminate STMs, (d) enhanced valid STMs.

explains why it is removed for lower available material amounts ($\beta = 0.1$) in the CTO results, since its contribution is rather low, albeit it is necessary for rigidity.

It should be noted that even indifferent STMs can be accepted within RC elements, as the given concrete volume and nonstructural reinforcements inherently ensure stability against minor deviations, cf. [45].

Example 5.3 (Corbel). The corbel attached to a column as shown in Figure 5.21a is loaded by a point load $F = 150$ kN. The FE mesh consists of $N_e = 40 \times 40 + 40 \times 100 = 5600$ quadrilateral elements (b). The boundary conditions are modeled in such a way that a STMs only for the corbel should evolve. Therefore, the upper column section remains unconstrained, whereas on the lower side one fixed support is placed at each corner. Neglecting the edge distances and concrete cover, their placement indicates the position of the resulting concrete compressive force and tensile force (rebars) in the column. This procedure meets necessary modeling requirements similar to those used in classical approaches to obtain STMs, cf. [2, 45]. Here, reference is also made to the modeling recommendations in Section 5.3.4.1.

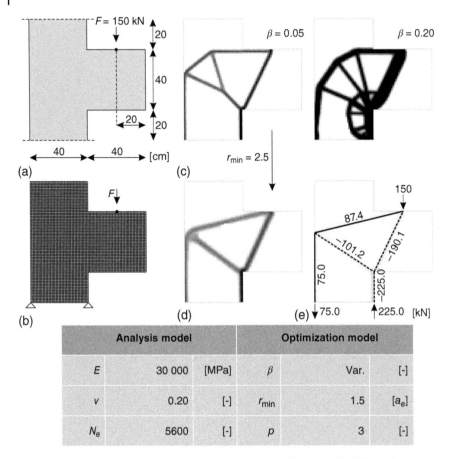

Figure 5.21 (a) Corbel problem, (b) analysis model, (c) CTO results, (d) CTO result with increased filter radius, (e) STMs.

Figure 5.21c shows the optimization results for $\beta = 0.05$ and $\beta = 0.20$ and a filter radius of $r_{min} = 1.5$ with respect to one element dimension. While the latter volume fraction limit leads to a highly braced structure, the result for the former is plain with distinct formation of the struts. Its direct transformation into a STMs, however, results in the STMs becoming kinematically indifferent. There are two ways to counteract this. The first is to replace the Y-shaped configuration of the inclined compression struts by a "V" layout in such a way that the compression node is connected to the upper tie via two straight compression struts. The second method is to increase the filter radius and thus force a denser material distribution within the design space. The result of this attempt with $r_{min} = 2.5$ is depicted in Figure 5.21d. It leads to a simple and – at least within the upper triangle – kinematically determined STMs (e). However, even indifferent model proposals can be accepted within a proper reinforced concrete structure.

Example 5.4 (Cantilever beam). The cantilever-like beam in Figure 5.22a is supported at the upper left-hand corner as well as at a distance $L/4$ from it. It is

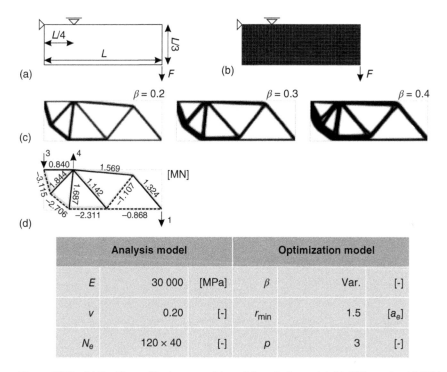

Figure 5.22 (a) Cantilever-like beam problem, (b) analysis model, (c) CTO results, (d) STMs.

	Analysis model			Optimization model	
E	30 000	[MPa]	β	Var.	[-]
v	0.20	[-]	r_{min}	1.5	$[a_e]$
N_e	120 × 40	[-]	p	3	[-]

loaded by a point load $F = 1$ MN at the lower right end. The analysis model is shown in Figure 5.22b where $N_e = 40 \times 120 = 4800$ equal quadrilateral elements are used for discretization.

The CTO results are shown in Figure 5.22c for different volume fractions. They all exhibit similar structures, where only the strut thicknesses differ to the same extent as the amount of available material increases. Deriving a STMs, therefore, proves to be straightforward. Figure 5.22d shows its member forces and support reactions relative to a load $F = 1$ MN.

Example 5.5 (Shear transfer at joints). Figure 5.23a shows the shear load transfer mechanism problem in segmented tunnel linings according to the cam & pot system. It transfers shear loads between adjacent rings in case of soft bedding ground. The "pot part" of the system is investigated here with a load F introduced at the corners of the opening. The design space is discretized with $N_e = 8161$ elements and the material properties are $E = 35\ 500$ MPa for the Young's modulus and $v = 0.2$ for the Poisson's ratio. The case has been studied in detail in [1, 46].

Figure 5.23b shows the CTO result for $\beta = 0.25$, which exhibits a material distribution with a distinct strut layout. The resulting STMs is shown in Figure 5.23c along with the emerging member forces due to a point load $F = 1$ MN. RC components developed and experimentally tested in [1, 46] with rebar layouts based on the optimization results exhibit up to 50 % higher load-bearing capacity and ductility in experiments compared to conventionally reinforced counterparts.

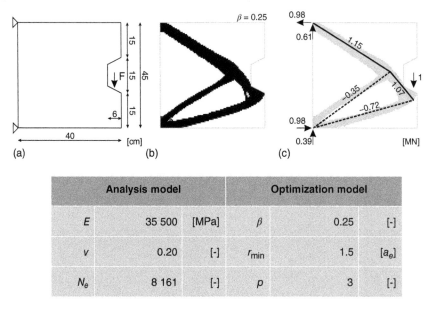

Analysis model			Optimization model		
E	35 500	[MPa]	β	0.25	[-]
v	0.20	[-]	r_{min}	1.5	$[a_e]$
N_e	8 161	[-]	p	3	[-]

Figure 5.23 (a) Shear load transfer problem, (b) CTO result, (c) STMs, adapted from [1].

5.5.2 TTO Approach

Example 5.6 (Deep beam 2). A simple deep beam corresponding to Example 5.1 with a span of $L = 400$ cm and a height of the same size as shown in Figure 5.24a is to be investigated, but, this time by means of TTO. The beam is loaded by a uniformly distributed load $q = 300$ kN/m applied to its upper edge. The distributed load is substituted for the analysis model by two resulting point loads: $F/2 = qL/2 = 300$ kN/m $\times 4.00$ m$/2 = 600$ kN. The layout of the base mesh must be oriented toward this, since a FE node needs to be located at each point of force application. For illustrative purposes, two different base meshes are examined: one with a block size of 4 $\times$ 4 and one with a block size of 8 $\times$ 8, cf. Figure 5.24b. The Young's modulus equals $E = 30\ 000$ MPa and the maximum cross-sectional area is $A_{max} = 100$ cm^2.

First, sample results for different constraint limits are calculated to determine a reasonable range of values for the constraint limit. Then, following the procedure described in Section 5.3.4.2, the STMs are refined by adjusting the permissible number of bars adequately and computing the optimization result therefore.

Using a minimum connectivity $Lvl = 1$, the sample results are shown in Figure 5.24c for a base mesh of 4×4 (c1) and 8×8 (c2) with limitation of the number of bars to 10, 20, and 30. For the coarser block size of 4×4, 10 and 20 bars prove to be too low. While 10 bars do not suffice to form a coherent structure, for 20 bars the design variables do not correspond to a full 0-1 distribution, which results in some members having a rather small cross-sectional area. With a constraint limit of $N_e^* = 30$, however, the algorithm converges to a complete structure with 0-1 distribution of the design variables. Further refinement is neither

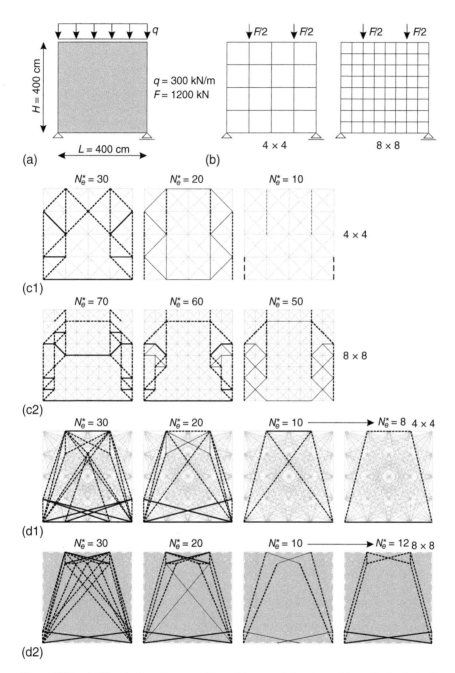

Figure 5.24 (a) Simple deep beam problem, (b) ground structures, (c) results for $Lvl = 1$, (d) optimization results for $Lvl = 4$ and $Lvl = 8$.

necessary nor possible to improve the result, but the STMs might be somewhat too sophisticated for practical application, since in total five tension ties have to be covered with reinforcement.

A base mesh of size 8×8 and $Lvl = 1$ leads to a result with similar conclusions. While a limitation to 50 bars must be considered too low, 60 and 70 permitted members lead to considerably complicated STMs showing fine 45 bracing and exhibiting numerous struts and ties, which all have to be designed at too great effort for practical applications.

Obviously, the low level of connectivity limits the potential load paths, regardless of the base mesh's size, to such an extent that full structures with sufficient number of bars tend to exhibit pronounced ramification in order to minimize the structural compliance.

Next, the results at maximum level of connectivity, $Lvl = 4$ and $Lvl = 8$ for the base meshes 4×4 (d1) and 8×8 (d2), respectively, are examined. Using the former, 30 and also 20 bars are too many and lead to complex structures. A constraint limit of $N_e^* = 10$ yields a simple structure, yet this cannot be used as STMs for design because of the crossing compression struts.[1] However, successive lowering of the constraint limit finally leads to a well-suited result at $N_e^* = 8$, which resembles the standard STMs for simple walls (Figure 5.25a,b).

The results for the base mesh of the size 8×8 lead to a similar conclusion as with the minimum level of connectivity. The least number of bars resulting in a full structure ($N_e^* = 12$) yields a result unsuitable for practical application due to the overlapping of compressive struts and structural nodes that are difficult to design. This gives an example of why it is important to build the base mesh following the guiding principle "as coarse as possible, as fine as necessary".

Consequently, the most suitable STMs for the design is achieved for a block size of 4×4, a maximum level of connectivity and the number of permissible bars limited to $N_e^* = 8$. Figure 5.25 provides a comparison of the computed STMs (c) with the standard model based on the elastic stress distribution of the uncracked wall (a) as well as with an extended model (b) that is more adequate for the ultimate limit state [45]. It can be seen that the STMs based on the stress distribution of the

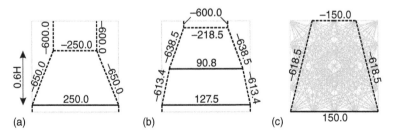

Figure 5.25 Axial forces [kN] of different STMs for a deep beam: (a) STMs oriented on the elastic stress distribution of the uncracked structure, (b) adjusted STMs for the ultimate limit state according to [45], (c) STMs resulting from the optimization.

---

1 Crossing compressive struts are physically impossible as they need to form a node of biaxial stresses, cf. [45].

uncracked state requires significantly more, namely 67 %, bending reinforcement than that resulting from the optimization (c). This observation is verified by using the extended model (b), which also requires significantly less bending tensile reinforcement at the lower edge. In practical applications, the steep vertical struts of case (c) – that small inclination causes the reduced tensile strut at the bottom – are usually omitted and replaced by minimum inclinations of 2:1 to prohibit extensive cracking [45]. This cracking will occur during the stress redistribution from the linear elastic to the "plastic" case of a STMs. According to [45], model (a) is primarily applicable to describe the serviceablity limit state, whereas in the ultimate limit state the compression zone constricts causing the lever arm between the resulting compression force and the lower bending reinforcement to increase significantly, which in turn reduces the lower horizontal tensile force resultant. Although the second tension tie (90.8 kN) is not represented in the computed STMs, the crack widths will be limited by the minimum surface reinforcement that has generally be installed as specified, for instance, by Schlaich et al. [2]. It should be noted that design approaches based on membrane theory, see for example [47], lead to reinforcement quantities that are similar to model (a).

Example 5.7 (Frame corner). The frame corner depicted in Figure 5.26a exposed to a negative bending moment M is to be designed. The height of the horizontal member corresponds to the height of the column's cross section and amounts to $h = 60$ cm. The bending moment equals $M = -150$ kN m and is represented by a tensile and compressive force resultant, F_t and F_c, respectively, for purposes of modeling. The inner lever arm is assumed to be $z \approx 0.8h$. From this follows for the resultants (internal forces): $F_t = F_c = M/z \approx M/(0.8h) = 150$ kN m$/(0.8 \times 0.6$ m$) = 312.5$ kN.

The design space for the optimization is limited by the position of the internal forces, which is assumed also for the column. For this reason, the supports are placed where the reinforcement (=the resultant tensile force) and the resultant compressive force are expected to lie, cf. Figure 5.26b. This ensures that a valid STMs can be computed. The base mesh is selected to be even and orthogonal. Both geometry and boundary conditions would also allow even larger distances between nodes, which, however, was not exploited within this example.

Figure 5.26c–d shows the TTO results for $Lvl = 1$ and $Lvl = Lvl_{max} = 9$. Starting with the latter (d), it is found that the resulting structures appear too complex for both a constraint limit of $N_e^* = 20$ and $N_e^* = 10$. For even lower numbers of allowable members, no coherent STMs with full cross-sectional areas is found. Thus, the maximum possible level of connectivity does not lead to acceptable results for design. By contrast, the sample results for the lowest possible level of connectivity, $Lvl = 1$, appear promising. Although limiting the number of bars to 20 and also 30 seems to be slightly too low, since members with small cross sections exist in the final structures, targeted adjustment of the constraint limit to $N_e^* = 32$ results in a practically sound STMs. In fact, it strongly resembles the established STMs according to [2] (Figure 5.26e), which underlines the validity of the optimization result.

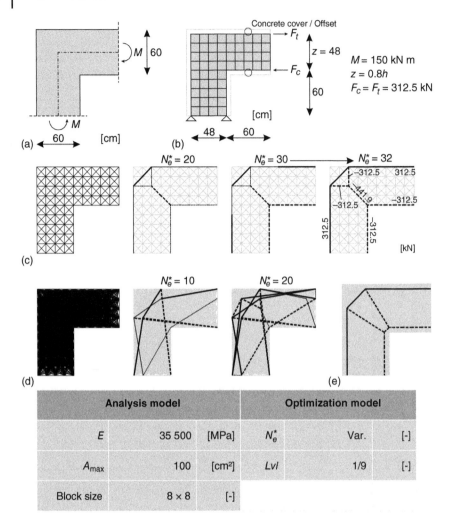

Figure 5.26 (a) Frame corner with negative bending moment, (b) model and base mesh, (c) results for $Lvl = 1$, (d) results for $Lvl = 9$, (e) classical STMs according to [2].

Example 5.8 (Wall with eccentric block-out). For the single-span deep beam with one large opening shown in Figure 5.27a, a STMs is sought for design. Its span length and height is $L = 700$ cm and $H = 450$ cm, respectively. The square opening is located in the lower left corner and measures 150 cm × 150 cm. The deep beam is subjected to a point load $F = 3000$ kN at a distance of 450 cm from the upper corner. The applied base mesh for TTO is displayed in Figure 5.27a. In the vicinity of the opening, the node distances are adaptively contracted and small, outside of it larger spacings are aimed for. The mesh is chosen to be orthogonal. Finally, the corresponding ground structures for $Lvl = 1$ and $Lvl = Lvl_{max} = 5$ are illustrated in Figure 5.27b.

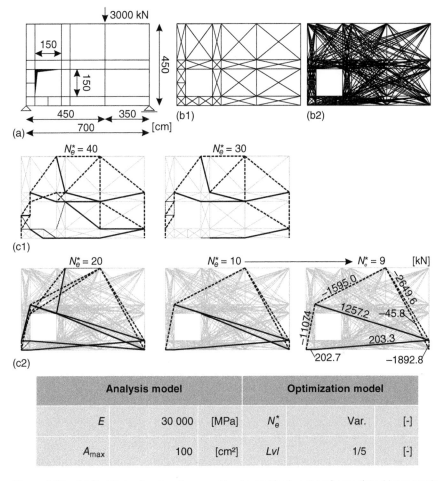

Figure 5.27 (a) Problem of a deep beam with a large block-out and associated base mesh, (b) ground structure for $Lvl = 1$ (b1) and $Lvl = Lvl_{max} = 5$ (b2), (c) optimization results for $Lvl = 1$ (c1) and $Lvl = 5$ (c2).

For $Lvl = 1$, the sample calculation for a constraint bound of $N_e^* = 40$ yields a rather complex structure (Figure 5.27c1). In contrast, the result for $N_e^* = 30$ permissible bars is incomplete, in particular the lower tensile member is not fully developed. Intermediate constraint limits lead to no significant improvements, namely to a complete but rather complex STMs. For the more strutted ground structure ($Lvl = Lvl_{max} = 5$, Figure 5.27c2), a constraint limit of $N_e^* = 20$ bars leads to a rather complex result, which is simplified for $N_e^* = 10$. It can be assumed, however, that further simplification might be feasible, which indeed is confirmed for $N_e^* = 9$, where two tensile members are merged into one.

Although the computed STMs appears to be well suited for design purposes, the problem mentioned in Section 5.3.4.1 of the tendency to favor inclined struts

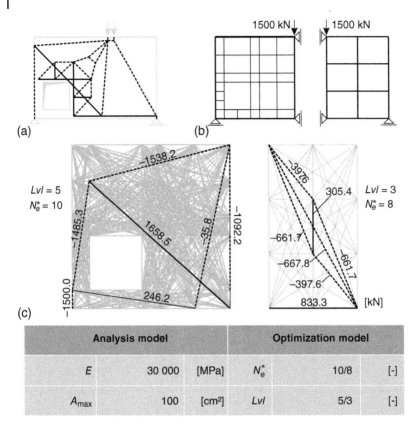

Figure 5.28 (a) Classical STMs for the deep beam with large opening problem according to [2], (b) base meshes for the subsystems, (c) optimization results for the subsystems.

instead of horizontal and vertical ones is evident in this example. While five bars would be required to form a lower horizontal tensile tie, the result with inclined strut requires merely one and is therefore preferred by the optimization algorithm. Nevertheless, in practice, rebars can be installed horizontally, using the horizontal force component for calculating the required cross-sectional area. It should also be noted that, compared to the established classical STMs (Figure 5.28a), the optimization result yields a less heavily reinforced region in the vicinity of the opening. This can lead to pronounced cracking, which should be counteracted by structurally reinforcing these areas. In addition, the compressive stresses of the two struts, which connect load application to the right-hand support (and could be merged into one), will spread bottle-shaped and thus cause additional lateral tensile stress [2]. The latter must also be structurally covered by reinforcement as is the case when applying the classical STMs. Typically, such additional reinforcements are covered by provided minimum surface mesh reinforcements.

Similar to [2], the STMs can be composed by calculating two subsystems, cf. Figure 5.28b. Doing so, two wall segments develop separated by horizontal bearings at the cut in the vertical level of expected horizontal struts. The vertical division is made at the position where the force is applied. This results in the left and right substitute system depicted in Figure 5.28b, each loaded by $F/2 = 1500$ kN. The section is located at the maximum bending moment, because there the shear force vanishes, and thus the principal stress trajectories orient just horizontally. In other words, the submodels are only connected by a horizontal compression and tension member. Consequently, placing the restraints corresponds to these boundary conditions.

As the optimization results for $Lvl = Lvl_{max}$ in Figure 5.28c reveal, this alternative approach has obvious advantages. By incorporating additional a priori information, the final STMs is improved in the sense that the optimization result approaches the classical STMs. The lower tie of the right subsystem is formed horizontally and turns upward only in the left subsystem. In addition, spreading of the compressive stress trajectories in the right-hand edge is approximately represented and the tensile strut there can be used to dimension the required lateral reinforcement.

Example 5.9 (Corbel with horizontal force). A STMs is sought for designing the corbel shown in Figure 5.29a. Length and height of the proceeding part both amount to 40 cm. The column, to which it is attached, has a thickness of the same magnitude. A horizontal force H acts concurrently with a vertical load F. F amounts to 150 kN and H amounts to 30 kN. A resultant load of $R = \sqrt{150^2 + 30^2} = 153$ kN develops with an inclination of $\tan^{-1}(1/5) = 11.3°$ to the vertical axis.

The analysis model is depicted in Figure 5.29b. When modeling the design space, both the concrete cover of the reinforcement and the edge distance of the compressive force resultant of the concrete are ignored for reasons of simplification within this example. The restraints are therefore placed punctually on the lower section of the column at the left and right edge nodes. Thus, it must be noted that the inner lever arm is slightly overestimated. No restraints are modeled at the upper section, since it can be assumed that the column is overpressed. Hence, the load transfer of the corbel is not prevailed by tension ties forming upward into the column. This is consistent with the classical STMs, which is shown in Figure 5.29c along with its resulting member forces. The upper tensile strut superimposes here from the share of F, deviated in 1:2 inclination, and H to $F/2 + H = 105$ kN. The base mesh comprises a square arrangement of the FE nodes at a distance of 10 cm.

For a minimum level of connectivity $Lvl = 1$, sample results are computed for a number of $N_e^* = 30$ and $N_e^* = 20$ permissible struts. The former constraint limit leads to an overloaded structure, the latter to an incomplete one with some struts exhibiting very small cross-sectional areas (Figure 5.29d1). Recalculation with an upper boundary of $N_e^* = 22$ members yields an improved result. The superfluous struts with particularly small cross sections can be neglected. The STMs is similar to the classical one in Figure 5.29c. However, the horizontal tensile chord is not formed at the corbel's upper edge but distributed over vertical levels, with diagonal struts

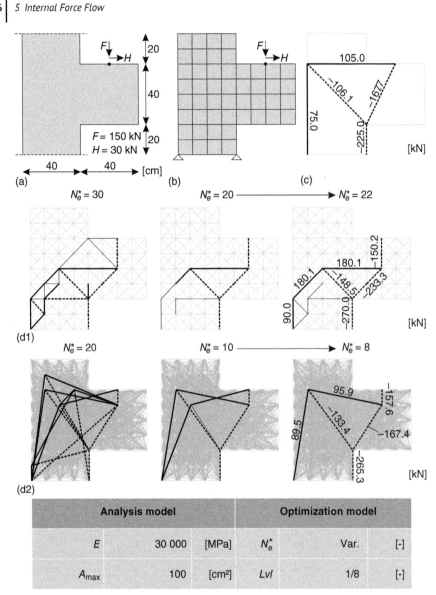

Figure 5.29 (a) Problem of a corbel, (b) analysis model and base mesh, (c) classical STMs according to [2], (d) optimization results for $Lvl = 1$ (d1) and $Lvl = 8$ (d2).

and a major tensile portion in the center. Due to the decreased lever arm, the tensile force is higher than with the former. The pronounced elongation of the vertical compressive strut, which starts from the load application and shifts the tensile tie downward, is attributed to the superiority of the vertical load component (F) relative to the horizontal one (H).

For $Lvl = Lvl_{max} = 8$, samples are computed for $N_e^* = 10$ and $N_e^* = 20$. The complexity of the result decreases gradually but is still slightly too pronounced for the latter (Figure 5.29d2). However, refining the structure by lowering the constraint limit in smaller steps leads to a well-suited STMs for $N_e^* = 8$. It is very similar to the classical STMs from [2]. Again, the tensile ties are inclined and not formed horizontally or vertically, as the latter demand multiple bars arranged in a row. The optimization algorithm therefore prefers the inclined, but single members. This results in a slightly, yet not significantly, different force distribution compared to the classical STMs in Figure 5.29c.

Example 5.10 (Stiffening core with openings). The wall structure in Figure 5.30a has a height of $H = 1500$ cm and a length of $L = 700$ cm. It exhibits

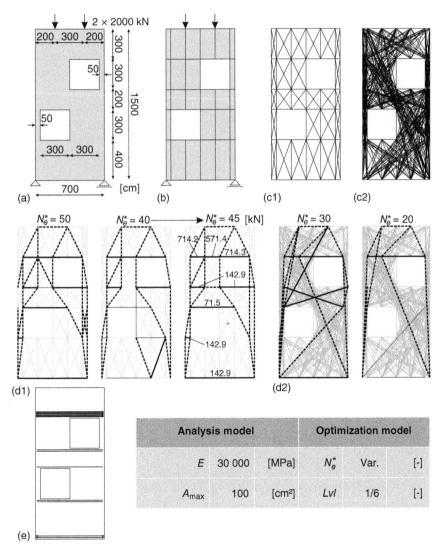

Analysis model			Optimization model		
E	30 000	[MPa]	N_e^*	Var.	[-]
A_{max}	100	[cm²]	Lvl	1/6	[-]

Figure 5.30 (a) Core wall with two openings problem, adapted from [48], (b) analysis model, (c) base mesh for $Lvl = 1$ (c1) and $Lvl = 6$ (c2), (d) optimization results for $Lvl = 1$ (d1) and $Lvl = 6$ (d2), (e) reinforcement layout.

two square openings that are arranged with a horizontal offset from each other and measure 300 cm × 300 cm each. The wall is subject to two point loads 2 × 2000 kN on its upper edge. The base mesh shown in Figure 5.30b is coarsely defined, albeit orthogonal, it is irregular due to the geometrical boundaries of the problem. Consequently, this applies likewise to the ground structures generated for $Lvl = 1$ and $Lvl = Lvl_{max} = 6$ in Figure 5.30c.

For a minimum level of connectivity ($Lvl = 1$), sample calculation is performed for $N_e^* = 40$ and $N_e^* = 50$ members (Figure 5.30d1). For the former, the structure is not completely formed and some struts are included which have only very small cross-sectional areas. On the other hand, the limitation to $N_e^* = 50$ leads to a result that appears slightly too complex. Furthermore, it is not suitable as STMs for practical application, since two compressive struts intersect overlapping on the right. Computing intermediate results by gradually adjusting the constraint limit within the interval]40,50[leads to a sound and valid STMs for $N_e^* = 45$. The resulting reinforcement layout is depicted in Figure 5.30e.

The situation is similar for the maximum level of connectivity $Lvl = 6$. While limiting the number of struts to $N_e^* = 20$ yields an incomplete structure containing members with small cross-sectional areas, $N_e^* = 30$ results in an unsuitable STMs with intersecting compressive struts. However, intermediate constraint limit values unfortunately do not lead to a STMs that is acceptable. Hence, the STMs resulting from $Lvl = 1$ and $N_e^* = 45$ members should be adopted for design.

Example 5.11 (Deep beam 3). A suitable design model is sought for the single-span beam in Figure 5.31a. The span length amounts to $L = 500$ cm and its height equals $H = 120$ cm. It has one small opening of the size 50 cm × 24 cm in the vicinity of each support. The beam is loaded by a vertical load $F = 140$ kN at midspan. The base mesh used is irregular due to the openings and consists of three block rows in total. The middle one contains units which are of equal size at the openings, while those above and below are somewhat larger.

First, samples for the lowest possible level of connectivity ($Lvl = 1$) are computed (Figure 5.31c1) based on the ground structure in Figure 5.31b1. Reasonable values for the constraint limit are confined to the interval]40,50[. While $N_e^* = 40$ proves to be too low, since struts with small cross-sectional areas remain, the resulting structure for $N_e^* = 50$ contains some superfluous members. The subsequent computations are found to eliminate these disadvantages by limiting the number of allowable bars to $N_e^* = 46$. However, due to the nature of the base mesh, struts and ties, which consists of multiple bars connected to each other, appear crooked. Moreover, their force deviation at the nodes causes the need for additional bracing.

The results for a minimum level of connectivity reveal the tendency of the load paths and hence the potential form of a reasonable STMs. Therefore, the second calculation approach is not performed with maximum level of connectivity. Instead, $Lvl = 3$ is chosen just to avoid crooked struts and ties. This makes it possible to span the upper and lower sides of the design space directly, using only one single strut. In addition, limiting the level of connectivity prevents overly complex structures to develop. Figure 5.31c2 presents the computed results. Through sampling,

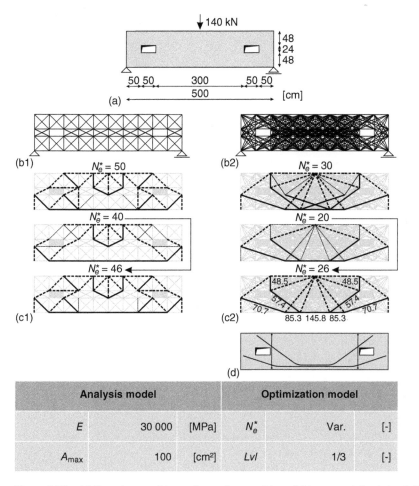

Figure 5.31 (a) Deep beam with small openings problem, (b) base mesh for $Lvl = 1$ (b1) and $Lvl = 3$ (b2), (c) optimization results for $Lvl = 1$ (c1) and $Lvl = 3$ (c2), (d) reinforcement layout.

a convenient constraint limit between $N_e^* = 20$ and $N_e^* = 30$ bars is specified. Finally, $N_e^* = 26$ leads to a STMs eligible for designing the reinforcement layout (d). Choosing $Lvl = 3$ proves adequate, since crooked struts and ties are avoided. Comparing the structural compliance ($c = 34.44$ kN cm) with that of the STMs for $Lvl = 1$ ($c = 67.53$ kN cm) also confirms that the latter is clearly inferior to the former.

Example 5.12 (Deep beam 4). A suitable STMs for the short deep beam with large holes in Figure 5.32a is to be computed. The span length of the beam equals $L = 800$ cm and its height to $H = 400$ cm. It is loaded by a vertical force $F = 140$ kN at midspan. The openings are rectangular and measure 200 cm × 100 cm each. The base mesh is orthogonal and the ground structures resulting for $Lvl = 1$ and $Lvl = Lvl_{max} = 8$ are shown in Figure 5.32b.

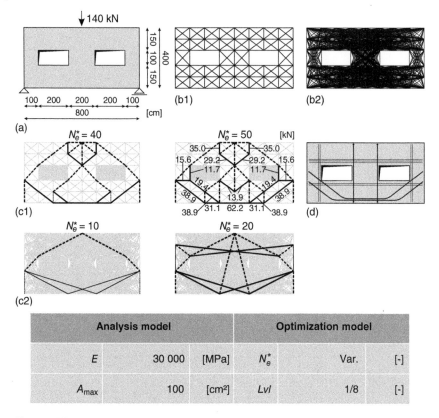

	Analysis model				Optimization model	
E	30 000	[MPa]	N_e^*		Var.	[-]
A_{max}	100	[cm²]	Lvl		1/8	[-]

Figure 5.32 (a) Short deep beam with large openings problem, (b) base mesh for $Lvl = 1$ (b1) and $Lvl = Lvl_{max} = 8$ (b2), (c) optimization results for $Lvl = 1$ (c1) and $Lvl = Lvl_{max} = 8$ (c2), (d) reinforcement layout.

The sampling results are shown in Figure 5.32c. While with $Lvl = 1$, for $N_e^* = 40$ bars the structure still contains members with small cross-sectional areas, a constraint value of $N_e^* = 50$ leads to a reasonable STMs. At maximum connectivity ($Lvl = 8$), $N_e^* = 10$ leads to an incomplete structure, while the result for $N_e^* = 20$ is rather complex. In addition, the load transfer appears cumbersome in the case of the latter. This is also confirmed by comparing its compliance ($c = 67.47$ kN cm) with that of the STMs for $Lvl = 1$ ($c = 34.04$ kN cm), which is almost twice as large. The most practical model for design is therefore that for $Lvl = 1$ and $N_e^* = 50$, to which the reinforcement layout in Figure 5.32d also refers.

Example 5.13 (Deep beam 5). A continuous deep beam spanning over two fields with a total of four openings is depicted in Figure 5.33a. A point load of the magnitude $F = 250$ kN acts at midspan of each field. The respective base mesh is given in Figure 5.33b.

For $Lvl = 1$, with a constraint limit of $N_e^* = 70$, the result appears too complex, while, in contrast, for $N_e^* = 60$ it is poorly developed. A constraint limit of $N_e^* = 65$ finally yields a reasonable STMs for designing the reinforcement layout (d).

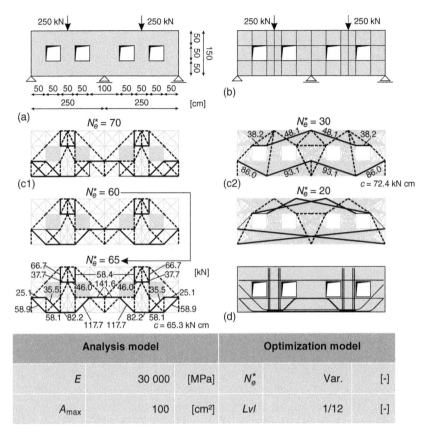

Figure 5.33 (a) Continuous deep beam with openings problem, (b) base mesh, (c) optimization results for $Lvl = 1$ (c1) and $Lvl = Lvl_{max} = 12$ (c2), (d) reinforcement layout.

On the other hand, at maximum level of connectivity ($Lvl = Lvl_{max} = 12$), $N_e^* = 20$ bars lead to a yet undeveloped result. In contrast, a good alternative STMs is calculated for $N_e^* = 30$, which shows fewer members and consequently implies a more simplified reinforcement layout than that resulting from $Lvl = 1$. The increased compliance value that has to be accepted in this case is tolerable and exceeds the minimum to approximately 11 %.

5.6 Applications

5.6.1 Joints in Tunnel Linings

Tunnels in mechanized tunneling usually consist of several segments forming lining rings. The lining rings are interconnected via circumferential joints, where shear loads are transferred, for example, by means of "cam & pot" system couplings as shown in Figure 5.34. Damages due to local shear peaks occur at these joints often during construction and thus reduce the tunnel's overall lifetime. The pot is thereby

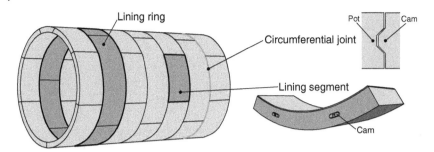

Figure 5.34 Components of a tunnel with "cam & pot" circumferential joints.

the most vulnerable part, since it typically can withstand only lower loads compared to the cam.

In [1, 46], one-material topology optimization is applied to determine convenient reinforcement layouts based on optimized STMs for the pot. The aim is to improve its robustness and load capacity, thus enhancing the overall durability of the tunnel. Figure 5.35 shows the computed optimized material distribution, its transfer to a STMs, the resulting reinforcement layout, and the practical realization in test specimens. A uniform mesh with a total of 7765 elements is employed for optimization. The assumed material parameters are $E = 35\,500$ MPa and $v = 0.2$, whereas the targeted volume fraction is $\beta = 0.25$.

The derived STMs is used to develop the reinforcement layout as well as the required cross-sectional areas of the rebars. The design concept proposes to transfer the load via inclined rebars to a welded steel plate which ensures anchoring of these. The rebar layout is verified in load tests in the laboratory on real scale specimens. The comparison with alternative design concepts reveals that the

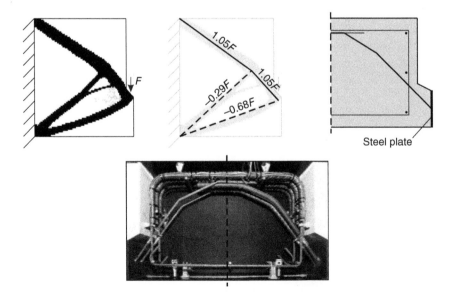

Figure 5.35 Optimization of the reinforcement layout at the pot of a circumferential joint.

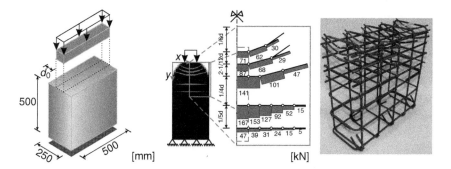

Figure 5.36 Test specimen representing the longitudinal joint of a tunnel lining segment, downscaled by 1:2, exemplary optimization result and reinforcement cage, adapted from [17].

optimized rebar layout not only ensures the highest load-bearing capacity but also significantly reduces scattering of the results. This demonstrates the superiority of the optimization aided approach both in the ultimate limit state and in terms of design reliability.

5.6.2 Partial Area Loading in Tunnel Linings

The presented CTTO approach was applied in [17] to longitudinal joints of segmental tunnel linings in order to optimize the reinforcement layout resulting from partial area loading. The geometry of the experimentally examined concrete block given in Figure 5.36 was extracted from a scaled down (1:2) tunnel lining segment. Within the optimization, centric and eccentric load application as well as multiple load cases were investigated to meet all real design requirements. From the result, several reinforcement concepts and their practical implementation were derived including conventional bar reinforcement, combinations of bars and fibers, fibers only configurations, and combinations of rebars with fiber cocktails (micro and macro fibers).

In some cases, moderate increases in load-bearing capacity were achieved in comparison to a conventional reinforcement concept. However, the most significant advantage of orienting the reinforcement layouts toward the optimization results was scattering reduction of the recorded ultimate loads within a test series, which significantly increases design safety and reliability.

References

1 Putke, T. (2016). Optimierungsgestützter Entwurf von Stahlbetonbauteilen am Beispiel von Tunnelschalen. PhD thesis. Bochum: Ruhr University Bochum.

2 Schlaich, J., Schäfer, K., and Jennewein, M. (1987). Toward a consistent design of structural concrete. *PCI Journal* 32 (3): 74–150.

3 Xia, Y., Langelaar, M., and Hendriks, M.A. (2020). A critical evaluation of topology optimization results for strut–and–tie modeling of reinforced concrete. *Computer-Aided Civil and Infrastructure Engineering* 35 (8): 850–869. https://doi.org/10.1111/mice.12537.

4 Dorn, W., Gomory, R.E., and Greenberg, H.J. (1964). Automatic design of optimal structures. *Journal de Mecanique* 3 (3): 25–52.

5 Fleron, P. (1964). The minimum weight of trusses. *Bygnings statiske Meddelelser* 35: 81–96.

6 Kirsch, U. (1989). Optimal topologies of structures. *Applied Mechanics Reviews* 42 (8): 223–239.

7 Pedersen, P. (1993). Topology optimization of three-dimensional trusses. In: *Topology Design of Structures, NATO ASI Series (Series E: Applied Sciences)*, Vol. 227, (eds. M.P. Bendsøe, C.A.M. Soares), Dordrecht: Springer. 19–30. https://doi.org/10.1007/978-94-011-1804-0_2.

8 Rozvany, G., Bendsøe, M.P., and Kirsch, U. (1995). Layout optimization of structures. *Applied Mechanics Reviews* 48 (2): 41–119. https://doi.org/10.1115/1.3101884.

9 Asadpoure, A., Guest, J.K., and Valdevit, L. (2015). Incorporating fabrication cost into topology optimization of discrete structures and lattices. *Structural and Multidisciplinary Optimization* 51: 385–396.

10 Bendsøe, M.P. (1989). Optimal shape design as a material distribution problem. *Structural Optimization* 1 (4): 193–202. https://doi.org/10.1007/BF01650949.

11 Bendsøe, M.P. and Sigmund, O. (1999). Material interpolation schemes in topology optimization. *Archive of Applied Mechanics* 69 (9–10): 635–654. https://doi.org/10.1007/s004190050248.

12 Bendsøe, M.P. and Sigmund, O. (2004). *Topology Optimization: Theory, Methods, and Applications*. Berlin: Springer-Verlag.

13 Mlejnek, H.-P. and Schirrmacher, R. (1993). An engineer's approach to optimal material distribution and shape finding. *Computer Methods in Applied Mechanics and Engineering* 106 (1–2): 1–26. https://doi.org/10.1016/0045-7825(93)90182-W.

14 Rozvany, G., Zhou, M., and Birker, T. (1992). Generalized shape optimization without homogenization. *Structural Optimization* 4 (3–4): 250–252. https://doi.org/10.1007/BF01742754.

15 Zhou, M. and Rozvany, G. (1991). The COC algorithm, Part II: Topological, geometrical and generalized shape optimization. *Computer Methods in Applied Mechanics and Engineering* 89 (1–3): 309–336. https://doi.org/10.1016/0045-7825(91)90046-9.

16 Zienkiewicz, O.C., Taylor, R.L., Nithiarasu, P., and Zhu, J.Z. (1977). *The Finite Element Method*, Vol. 3. London: McGraw-Hill.

17 Smarslik, M. (2019). Optimization-based design of structural concrete using hybrid reinforcements. PhD thesis. Bochum: Ruhr University Bochum.

18 Amir, O. (2013). A topology optimization procedure for reinforced concrete structures. *Computers & Structures* 114–115: 46–58.

19 He, L. and Gilbert, M. (2015). Rationalization of trusses generated via layout optimization. *Structural and Multidisciplinary Optimization* 52 (4): 677–694.

20 Parkes, E.W. (1975). Joints in optimum frameworks. *International Journal of Solids and Structures* 11 (9): 1017–1022.

21 Torii, A.J., Lopez, R.H., and Miguel, L.F.F. (2016). Design complexity control in truss optimization. *Structural and Multidisciplinary Optimization* 54 (2): 289–299.

22 Zegard, T. and Paulino, G.H. (2014). GRAND — Ground structure based topology optimization for arbitrary 2D domains using MATLAB. *Structural and Multidisciplinary Optimization* 50 (5): 861–882. https://doi.org/10.1007/s00158-014-1085-z.

23 Svanberg, K. (1987). The method of moving asymptotes: a new method for structural optimization. *International Journal for Numerical Methods in Engineering* 24 (2): 359–373. https://doi.org/10.1002/nme.1620240207.

24 Harzheim, L. (2008). *Strukturoptimierung: Grundlagen und Anwendungen.* Frankfurt am Main: Harri Deutsch.

25 Haftka, R.T. and Gürdal, Z. (1993). *Elements of Structural Optimization.* Dordrecht: Kluwer.

26 Sigmund, O. (2001). A 99 line topology optimization code written in Matlab. *Structural and Multidisciplinary Optimization* 21 (2): 120–127.

27 Andreassen, E., Clausen, A., Schevenels, M. et al. (2011). Efficient topology optimization in MATLAB using 88 lines of code. *Structural and Multidisciplinary Optimization* 43 (1): 1–16.

28 Bathe, K.-J. (1996). *Finite Element Procedures, Prentice Hall International Editions.* Englewood Cliffs, NJ: Prentice Hall. ISBN: 013349697x.

29 Hughes, T.J.R. (2000). *The Finite Element Method: Linear Static and Dynamic Finite Element Analysis.* Mineola, NY: Dover, Reprint ISBN: 0486411818. http://www.loc.gov/catdir/description/dover031/00038414.html.

30 Amir, O. and Sigmund, O. (2013). Reinforcement layout design for concrete structures based on continuum damage and truss topology optimization. *Structural and Multidisciplinary Optimization* 47 (2): 157–174.

31 Bogomolny, M. and Amir, O. (2012). Conceptual design of reinforced concrete structures using topology optimization with elastoplastic material modeling. *International Journal for Numerical Methods in Engineering* 90 (13): 1578–1597. https://doi.org/10.1002/nme.4253.

32 Drucker, D.C. and Prager, W. (1952). Soil mechanics and plastic analysis or limit design. *Quaterly of Applied Mathematics* 10 (2): 157–165.

33 Gaynor, A.T., Guest, J.K., and Moen, C.D. (2013). Reinforced concrete force visualization and design using bilinear truss-continuum topology optimization. *Journal of Structural Engineering* 139 (4): 607–618. https://doi.org/10.1061/(ASCE)ST.1943-541X.0000692.

34 Yang, Y., Moen, C.D., and Guest, J. (2013). Computer-generated force flow paths for concrete design: An alternative to traditional strut-and-tie models. *Convention and National Bridge Conference*, Grapevine, TX, USA.

35 Yang, Y., Guest, J., and Moen, C.D. (2014). Optimizing reinforcement layout in concrete design considering constructability. *2014 Convention and National Bridge Conference*, Washingtion, D.C., USA.

36 Smarslik, M.J., Neu, G., Mark, P. et al. (2017). Robust reinforcement layout for segmental tunnel lining under partial area loading using hybrid topology optimization. In: *Computational Methods in Tunneling and Subsurface Engineering (EURO:TUN 2017)* (eds. G. Hofstetter et al.), 321–328. Universität Inssbruck.

37 Mark, P., Smarslik, M.J., and Tabka, B. (2018). Optimized reinforcement for partial area loading. *BFT International* 84 (2): 61.

38 Smarslik, M. and Mark, P. (2019). Hybrid reinforcement design of longitudinal joints for segmental concrete linings. *Structural Concrete* 20 (6): 1926–1940.

39 Darwin, D. and Pecknold, D.A. (1974). Inealstic Model for Cyclic Biaxial Loading of Reinforced Concrete. *Structural Research Series 409*.Urbana, IL: University of Illinois.

40 Darwin, D. and Pecknold, D.A. (1977). Nonlinear biaxial stress-strain law for concrete. *Journal of the Engineering Mechanics Division* 103 (4): 229–241.

41 Sigmund, O. (2007). Morphology-based black and white filters for topology optimization. *Structural and Multidisciplinary Optimization* 33 (4–5): 401–424. https://doi.org/10.1007/s00158-006-0087-x.

42 Sigmund, O. and Petersson, J. (1998). Numerical instabilities in topology optimization: A survey on procedures dealing with checkerboards, mesh-dependencies and local minima. *Structural Optimization* 16 (1): 68–75. https://doi.org/10.1007/BF01214002.

43 Aurenhammer, F., Klein, R., and Lee, D.-T. (2013). *Voronoi Diagrams and Delaunay Triangulations*. New Jersey, USA: World Scientific.

44 Okabe, A., Boots, B.N., Sugihara, K. et al. (2008). *Spatial Tessallations: Concepts and Applications of Voronoi Diagrams*. New York: Wiley. ISBN: 9780471986355.

45 Schlaich, J. and Schäfer, K. (2001). *Konstruieren im Stahlbetonbau: Beton-Kalender 2001*, 311–492. Berlin: Ernst & Sohn.

46 Putke, T., Bohun, R., and Mark, P. (2015). Experimental analyses of an optimized shear load transfer in the circumferential joints of concrete segmental linings. *Structural Concrete* 16 (4): 572–582. https://doi.org/10.1002/suco .201500013.

47 DAfStb-Heft 631 (2019). *Hilfsmittel zur Schnittgrößenermittlung und zu besonderen Detailnachweisen bei Stahlbetontragwerken*, Beuth-Verlag.

48 Reineck, K.-H. (ed.) (2002). *Examples for the Design of Structural Concrete with Strut-and-Tie Models, ACI International SP*, vol. 208. Farmington Hills, MI: ACI International. ISBN: 0870310860.

6

Design of Cross-sections

Key learnings after reading this chapter:

- How can the shape and material distribution, i.e. concrete and reinforcement, of a cross-section be optimally adapted to actions?
- How can cross-sectional classes be parameterized for the optimization?
- How can environmental sustainability, costs, and other assessment criteria be integrated into the cross-sectional design?

This chapter deals with the optimization aided design of composite cross-sections of arbitrary material combinations. Emphasis is placed on bending design both with and without normal force for plane and spatial situations.

Starting point is the common engineering practice. In this process, a cross-sectional type with its dimensions is first chosen by intuition. Subsequently, the strain plane leading to equilibrium between internal and external forces is determined in an iterative manner. In the case of RC, for instance, the required reinforcement quantities are then computed accordingly. However, two issues emerge with this approach. First, the iterative determination of the strain plane can become very cumbersome, particularly for spatial loading conditions and asymmetric cross-sections or material distributions. Second, the cross-sectional dimensions are preliminarily specified by intuition rather than being directly obtained from objective and verifiable criteria.

In order to circumvent the above-mentioned shortcomings, suitable optimization methods are presented. Specifically, first, an approach is presented to determine the strain plane for arbitrary composite cross-sections for plane and spatial loading conditions. Subsequently, the method is extended to be applicable to more sophisticated problems, namely, to determine the optimal cross-sectional shape while complying with the material stress limits. Depending on the particular requirements, the objective to be minimized can be of various types, for instance, the amount of formwork used or the overall monetary and environmental structural costs. A key aspect is the convenient parametrization of cross-sectional types, meaning their discretization by

(Continued)

Optimization Aided Design: Reinforced Concrete, First Edition.
Georgios Gaganelis, Peter Mark, and Patrick Forman.
© 2022 Ernst & Sohn GmbH & Co.KG. Published 2022 by Ernst & Sohn GmbH & Co.KG.

(Continued)

subareas, so-called lamellae or fibers (cf. Figure 6.1), in order to ensure easy applicability to a wide range of settings.

The benefits of applying optimization methods to the design of cross-sections are versatile, such as ideally adapted shapes to defined criteria, enhanced transparency of design decisions, improved cost efficiency, and significant reduction of time needed to solve the design task, even for spatially or geometrically complicated cross-sections.

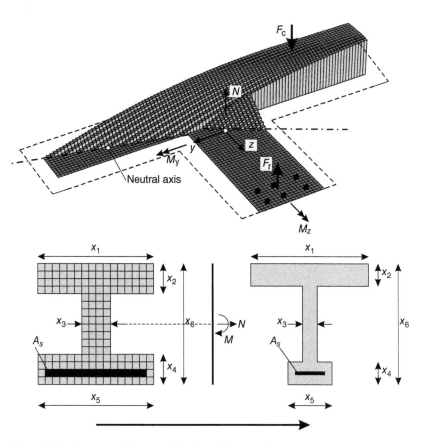

Figure 6.1 Overview of Chapter 6: design of Cross-sections.

6.1   Problem Statement

The cross-sectional design constitutes a common task of the design in ultimate limit state (ULS) [1–3]. Thereby, maximum stress and strain values of the materials must generally be respected. However, examination at the cross-sectional level is also important in serviceability limit state (SLS), for example when checking impermeability (sufficient compression zone height), decompression, and stresses. This chapter focuses on optimization aided methods for a bending design including axial forces, if present, cf. Section 2.5.1. Shear and torsion design are not covered. Doing so, only axial stresses σ are treated.

The choice of a basic cross-sectional shape is so far made intuitively and considered as default for the bending design. This concerns all materials, the cross-sectional type, as well as its heights, widths, and other dimensions. The design always refers exclusively to dimensioning of one single quantity, for instance, the required amount of reinforcement. In doing so, other parameters like shapes, material strengths, or basic configurations of cross-sections remain fixed and uncontrolled for effectivity. The actual design of a cross-sectional layout is thus disregarded, that is, answering the questions of what shape, material distribution, strengths, and other quantities are most convenient for the given actions. The answer just proves sufficient load-bearing capacity or serviceability. Whether the cross-section is well designed in materials, geometric parameters, and locations of reinforcements in a sense of really minimizing overall efforts is not treated.

The aim must rather be to find a holistic cross-sectional design that complies with boundary conditions predefined by the user, for example, the least possible amount of formwork, minimum cross-sectional area, minimal use of reinforcements or combinations of these criteria. In doing so, verifications of structural capacity or serviceability should be inherently achieved. Dimensioning of reinforcement thus becomes a side issue. The result is a shift from an intuition-based design approach to one based on distinct criteria. Moreover, other specifications such as manufacturing, durability (concrete cover), structural fire protection, robustness (impact), and aspiring to simple geometries for the building process must also be taken into account. Such criteria can be included in the form of constraints within the optimization.

Cross-sections can be arbitrary and made of different materials. The cross-sectional plane is defined by the y and z coordinates, whereas x denotes the longitudinal axis, see Figure 6.2. Generally, no limitations exist concerning their types, quantities, and distributions as well as on cross-sectional shape and spatial orientations. For the basic assumptions of the cross-sectional design, the reader is

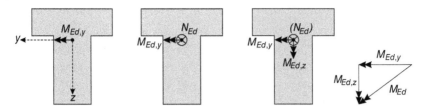

Figure 6.2 Combination of sectional forces: uniaxial bending, uniaxial bending with axial force, and biaxial bending.

referred to Section 2.5.1, where all basic equations are derived. Summarizing their content, they are repeated like:

- Sections are loaded by sectional forces of bending M_{Ed} and axial forces N_{Ed} (plane bending), or $M_{Ed,y}$, $M_{Ed,z}$ and N_{Ed} in the spatial case of biaxial bending. The subscript "d" indicates design values, while "E" describes effects of actions. Sectional forces are treated as input variables.
- Plane sections are supposed to remain plane. Thus, strains ε form a plane that can be described by axial strains and curvatures, denoted here by b_1, b_2, and b_3. b_3 arises in a spatial case.
- For all i materials used in a cross-section, stress–strain relations $\sigma_i = \sigma_i(\varepsilon)$ are defined. They can have linear or nonlinear courses as long as they stay physically reasonable.
- The bond between all materials is assumed to be free of slip. So the strain at positions y and z in a cross-section directly yield corresponding stresses.
- The stresses of the cross-section are integrated for sectional forces. The equilibrium conditions – two in the plane case, three in the spatial one – ensure that sectional forces of loadings (N_{Ed}, M_{Ed} in plane case; N_{Ed}, $M_{Ed,y}$, $M_{Ed,z}$ in spatial case) coincide with corresponding stress integrals.

The design question of finding stresses, suitable minimum reinforcement amounts, or geometric parameters out of given sectional forces demands iterative solutions, as the strain plane has to be moved and rotated with respect to axial strains and curvatures, so b_1, b_2, b_3.

Due to the typical design approach described above, the adequacy of the choices remains unclear, so mostly simple cross-sectional types made of few materials and with basic shapes are usually chosen. Typical examples are cross-sections made of RC, fiber-reinforced concrete (FRC), hybrid reinforcements of steel fibers and RC [4–6], composite cross-sections of sectional steel with RC filling (hollow cross-sections) [7, 8] or casing (open cross-sections with lining), as described in Figure 6.3. They are usually solid (plates, rectangular, and circular) and exhibit polygonal shapes. More advanced designs employing material-saving principles, adopt I- or T-shapes with compression and tension zones made of concrete and steel, respectively. Single or double symmetrical cross-sections are typical. However, in case of one-sided support, asymmetrical geometries, such as L-shapes, may also be found. Actions can be axial forces N_{Ed}, as well as bending moments $M_{Ed} = M_{Ed,y}$ and $M_{Ed} = \left[M_{Ed,y}, M_{Ed,z}\right]$ in the plane and spatial case, respectively. Common

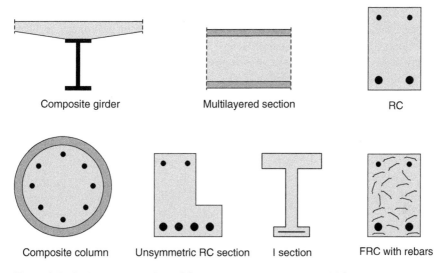

Composite girder Multilayered section RC

Composite column Unsymmetric RC section I section FRC with rebars

Figure 6.3 Typical cross-sections: RC, steel–concrete composites, FRC, layered sections.

scenarios, in practice, are uniaxial bending (plane beams), uniaxial bending with normal force (plane frames, bars, and beams), or biaxial bending with or without normal force (spatial beams).

For a given combination of actions, the resulting axial stress distribution σ within the cross-section is sought by satisfying equilibrium. This corresponds to the typical design situation outlined in Section 2.5.1, namely identifying the internal forces and computing the resulting strains and stresses for the cross-section. For given internal forces and fixed geometry, the problem is solvable, although generally only via an iterative approach due to the typically nonlinear constitutive equations. Relevant equilibrium states for dimensioning appear on the boundaries of the solution space. In RC design, this is always the case when either the outer edge of the concrete reaches its minimum compressive strain ε_{cu} or when the reinforcing steel meets its maximum tensile strain ε_{su}. The fundamental idea of optimization becomes obvious at this point: Since the design follows the requirement of cost efficiency, a minimization problem is set up. For example, the task could be to determine the minimum amount of reinforcement in order to ensure load-bearing capacity. Different further objectives and restrictions are also conceivable, such as the least possible amount of material, lowest height, and minimal formwork effort with respect to boundary constraints, for instance, minimum concrete cover, maximum amount of reinforcement, and maximum heights or minimum widths.

6.2 Equilibrium Iteration

Related Examples: 6.1–6.3.

Assume that the internal forces N and M_y in the plane case and N, M_y, and M_z in the spatial case are given. For the sake of simplicity, subscripts "d" and "E" are neglected. Provided that, in addition, the cross-section is also predefined including complete information on geometry, materials, and reinforcement, dimensioning of these quantities is not performed. Instead, the optimization problem is reduced to merely finding a strain and corresponding stress distribution. In other words, the goal is to effectively solve a system of nonlinear equations. In doing so, only one distinct solution exists. It is found in a design case when boundaries are met, such as maximum or minimum strain and stress values. This procedure corresponds to the basic concept of RC design aids for the ULS, where also only the equilibrium conditions are solved [9, 10], however, by utilizing specifical limit strain values in order to thus obtain the minimum required reinforcement amounts.

The design variables in the optimization are defined as the parameters b_1, b_2, and b_3 describing the strain plane according to Eq. (2.4) (plane) or 2.5 (spatial):

$$\mathbf{x} = \begin{bmatrix} b_1 & b_2 \end{bmatrix}^\mathsf{T} \tag{6.1a}$$

$$\mathbf{x} = \begin{bmatrix} b_1 & b_2 & b_3 \end{bmatrix}^\mathsf{T} \tag{6.1b}$$

As objective function serves the minimization of the error sums derived from the equilibrium equations [11]. For the plane case, it holds

$$f_{2D} = \left(N - \int_A \sigma dA \right)^p + \left(M_y - \int_A \sigma z dA \right)^p \rightarrow \min \tag{6.2}$$

and for the spatial case

$$f_{3D} = \left(N - \int_A \sigma dA \right)^p + \left(M_y - \int_A \sigma z dA \right)^p + \left(M_z + \int_A \sigma y dA \right)^p \rightarrow \min \tag{6.3}$$

An even number is adopted as exponent p, where for $p = 2$ the equations correspond to minimizing error squares. Generally, $p = 2$ or $p = 4$ is recommended for practical application [12]. If equilibrium prevails, the respective actions are equal to the corresponding stress integrals within the brackets and the error squares disappear. This leads to the objective function becoming zero. Minimizing the error sum thus simultaneously solves the equilibrium. Furthermore, a criterion exists, which allows to control the found solution: global minimum is reached, if $f = 0$, meaning that the individual actions correspond to the respective integrals. Otherwise, the optimization must be repeated with different initial design variable values.

Equilibrium is only achieved if the cross-section provides sufficient load-bearing capacity. Therefore, it is not worth defining limits for stresses and strains, since only one solution is possible and introducing additional boundary conditions is redundant.

The individual terms of the objective function related to N, M_y, and M_z should not differ too much in magnitude in order to yield a comparable proportion to the objective function value. If the numerical differences between N, M_y, and M_z are too large, scaling or assigning a prefactor to the smallest summand may help. Nontrivial initial values in $\mathbf{x}$ for b_1, b_2, and b_3, which are not too far away from the actual solution, are preferable. A simple linear elastic estimation or intuitive trial and error will provide good guidance. Zero vectors should be avoided in any case as initial configuration, because this may cause the solver to fail in finding a nontrivial solution.

6.3 Sectional Optimization

Related Examples: 6.4–6.6.

The optimization can be set up in such a way that both the cross-sectional dimensions are optimized and equilibrium is found, where the stress and strain limits are met. In doing so, the actual ULS design becomes a constraint to be fulfilled. In mathematical terms, this entails achieving equilibrium (equality constraint) and meeting stress or strain limits (inequality constraints), while minimizing specific cross-sectional quantities. The design variables, such as geometries, reinforcement amounts, and strain parameters, must respect these constraints. Some examples are maximum side lengths, reinforcement quantities must be nonzero to maintain physical plausibility, and maximum lengths to be maintained in order that the reinforcement fits within the cross-section. The cross-sectional adaptation (design) is included within the objective function.

The design variables are specified in the vector $\mathbf{x}$:

$$\mathbf{x} = \begin{bmatrix} b_1 & b_2 & x_1 & x_2 & ... \end{bmatrix}^\mathsf{T} \tag{6.4a}$$

$$\mathbf{x} = \begin{bmatrix} b_1 & b_2 & b_3 & x_1 & x_2 & ... \end{bmatrix}^\mathsf{T} \tag{6.4b}$$

for the plane and spatial case, respectively. They include the parameters describing the strain plane (b_1, b_2 and, if applicable, b_3) in order to specify the equilibrium state. Further variables x_i may be of geometrical or material nature, depending on the task. Typical examples for design variables x_i are reinforcement quantities A_{si}, cross-sectional dimensions such as side lengths, thicknesses, positions of reinforcements, and openings, as well as material strengths and their distribution within the cross-section. It should be noted that, in order to optimize the shape of the latter, at least one variable x_1 has to be defined.

Generally, a conflict exists between the number of design variables x_i and the robustness of convergence. As a rule, the more the design variables are defined, the more freely the shape of the cross-section can be optimized. At the same time, however, the sensitivity of each design variable on the objective function decreases and multiple solutions may become feasible. The transparency for the user reduces and the convergence behavior is disturbed such that, in the worst case, even finding a solution is inhibited. In contrast, a small number of design variables favors convergence behavior and comprehensibility of the problem but may prevent finding innovative nontrivial solutions due to the constraint on the solution space.

In the optimization aided design of the cross-section, the equilibrium is formulated as a set of equality constraints:

$$h_1(\mathbf{x}) = N - \int_A \sigma(\mathbf{x})dA \overset{!}{=} 0 \tag{6.5a}$$

$$h_2(\mathbf{x}) = M_y - \int_A \sigma(\mathbf{x})zdA \overset{!}{=} 0 \tag{6.5b}$$

for the plane and

$$h_1(\mathbf{x}) = N - \int_A \sigma(\mathbf{x})dA \overset{!}{=} 0 \qquad (6.6a)$$

$$h_2(\mathbf{x}) = M_y - \int_A \sigma(\mathbf{x})zdA \overset{!}{=} 0 \qquad (6.6b)$$

$$h_3(\mathbf{x}) = M_z + \int_A \sigma(\mathbf{x})ydA \overset{!}{=} 0 \qquad (6.6c)$$

for the spatial case. The stress integrals depend on the strain state (b_1, b_2 and, if applicable, b_3), as well as the additional design variables ($x_1, x_2, ...$), hence generally from $\mathbf{x}$.

For the reasons of physical sufficiency, the design variables must be limited through inequality constraints within reasonable lower (superscript L) and upper (superscript U) limits:

$$\begin{aligned} x_1^L &= 0 \le x_1 \le x_1^U \\ x_2^L &= 0 \le x_2 \le x_2^U \\ &\vdots \end{aligned} \qquad (6.7)$$

The lower boundary is at least zero, since physical quantities such as lengths and material strengths must not become negative. In addition, quantities like geometric dimensions, strength values, and areas are often physically limited by upper limits. Examples are maximum possible concrete strength classes or maximum cross-sectional heights due to a clearance that must be complied with.

In order to limit the stress and the strain values σ and ε, respectively, additional inequality constraints must be included in the formulation of the optimization problem, namely:

$$g_1(\mathbf{x}) = \sigma^L - \sigma(\mathbf{x}, y, z) \le 0 \qquad (6.8a)$$

$$g_2(\mathbf{x}) = \sigma(\mathbf{x}, y, z) - \sigma^U \le 0 \qquad (6.8b)$$

and

$$g_1(\mathbf{x}) = \varepsilon^L - \varepsilon(\mathbf{x}, y, z) \le 0 \qquad (6.9a)$$

$$g_2(\mathbf{x}) = \varepsilon(\mathbf{x}, y, z) - \varepsilon^U \le 0 \qquad (6.9b)$$

Such limits are to be checked for each material (M_1, M_2, ...), thus within the respective areas (A_1, A_2, ...), for all related y- and z-coordinates. However, due to the underlying strain plane, checking is usually sufficient at the boundaries (or even only at the vertices) of the individual areas within the cross-section. In doing so, meeting the stress and strain limits provides evidence that statically feasible solutions have been found (ULS design).

Furthermore, geometric limits, such as the maximum total reinforcement amount $A_{s,tot}$, which generally is the sum of several single quantities, may also become relevant and must therefore be constrained, for instance via

$$g(\mathbf{x}) = A_{s1} + A_{s2} + A_{s3} + \cdots \le A_s^U \qquad (6.10)$$

In the same way, additional inequality constraints can be incorporated, if needed.

The objective is the mathematical formulation of a leading design ideal expressed as a minimization task. Usually, this task is to minimize the overall costs. It is, therefore, necessary to link the cross-sectional geometric or material quantities with such cost values. Numerous formulations are both possible and reasonable. In the following, some typical examples are provided which are generally applicable to problems in practice.

6.3.1 Reinforcement Amounts

The cross-section should have a specified shape and given material parameters for concrete and reinforcement. Then, the longitudinal reinforcement quantities to be installed at preselected positions are sought. The total reinforcement $A_{s,\text{tot}}$, which results as the sum of the individual quantities ($x_1 = A_1, x_2 = A_2, \ldots$), is to be minimized:

$$f(\mathbf{x}) = \sum_i A_{si} \rightarrow \min \tag{6.11}$$

In doing so, the costs are equated with the total amount of reinforcement. The reinforcement amounts are treated as free optimization variables.

6.3.2 Cross-sectional Layout

The cross-section is defined by different design parameters x_i, representing physical dimensions (lengths or thicknesses) which have to be adjusted within the optimization while complying with stress or strain limits in such a way that the total cross-sectional area is minimized. In this case, the costs correspond to the material consumption. The total cross-sectional area A is thus to be minimized:

$$f(\mathbf{x}) = A \rightarrow \min \tag{6.12}$$

The design variables $\mathbf{x}$ describe A by geometric parameters. Shuttering work due to, for instance, complicated geometries can also be incorporated into the costs (minimization of edge lengths).

The objective function may be inhomogeneous, i.e. it can consist of quantities with different units, but only in such a way that it provides sufficient sensitivity to the optimization variables x_i. However, it is recommended to ensure compliant units, for example, monetary costs, which arise from the individual parameters by a proper relation. In this way, it is possible to conveniently combine quantities that are physically different (lengths, areas, and strengths) but generate costs via comparative cost values (e.g. € /m, € /m², and € /(MN/m²)).

6.3.3 Material Weighting

The use of different materials is often associated with different expenses. The term "expenses" may refer to monetary costs or, for example, to alternative costs such as the material-related carbon footprint. RC is a typical example, where a ton of reinforcing steel is significantly more expensive than a ton of concrete, both in terms of price and environmental impact. At the cross-sectional level, these different costs can be accounted for by using weighting factors α_j for the individual material areas A_j. A high weighting factor thus makes the use of a material disadvantageous, while a lower one favors it accordingly. Following this concept, the objective function reads

$$f(\mathbf{x}) = \sum_j \alpha_j A_j \rightarrow \min \tag{6.13}$$

Conceptually, Eq. (6.13) corresponds to minimizing the cross-sectional area in the same way as Eq. (6.12), however, now differentiated with respect to the various materials incorporated. The describing parameters to specify A_j are used as optimization variables x.

6.4 Solving

6.4.1 Stress Integrals

The stress integrals of the equilibrium from Eqs. (2.9) and (2.10) are evaluated at each optimization step. Numerical integration can be used for complex cross-sections or elaborate constitutive equations [12–15], but also using closed-form solutions [5] for more simple material equations and geometries is possible.

In case of the numerical approach, it is useful to discretize the cross-section similar to finite elements into partial areas ΔA in which the integrals are solved by approximating the stress distribution using simplified methods. Generally, it is straightforward to apply fine subdivisions (principle of h-adaptation), thus to specify a large number of subareas ΔA, allowing to obtain very accurate results even with rather simple stress approximations. By integrating approximated stress curves over many discrete areas, inaccuracies are equalized and do hardly affect the final quantities of the computed forces or moments from the individual materials. Higher-order interpolations (principle of p-adaptation), for instance, Gaussian or Newton–Cotes quadrature, do not offer further advantages [15, 16].

Assuming a constant stress σ_i within a subarea ΔA_i (stresses at the centroid of ΔA_i are applied all over the subarea), the stress integrals transform into easy to calculate summations. In doing so, each material M_i with its respective area A_i and associated stress–strain relation $\sigma_i(\varepsilon)$ is treated separately and discretized into subelements ΔA_i. Following this approach, the stress integrals from Eq. (2.10) yield

$$\int_{A_1} \sigma_1 dA_1 + \int_{A_2} \sigma_2 dA_2 + \cdots \approx \sum_i \sigma_{1i}\Delta A_{1i} + \sum_j \sigma_{2j}\Delta A_{2j} + \cdots \tag{6.14a}$$

$$\int_{A_1} \sigma_1 z dA_1 + \int_{A_2} \sigma_2 z dA_2 + \cdots \approx \sum_i \sigma_{1i} z_i \Delta A_{1i} + \sum_j \sigma_{2j} z_j \Delta A_{2j} + \cdots \tag{6.14b}$$

$$-\int_{A_1} \sigma_1 y dA_1 - \int_{A_2} \sigma_2 y dA_2 + \cdots \approx -\sum_i \sigma_{1i} y_i \Delta A_{1i} - \sum_j \sigma_{2j} y_j \Delta A_{2j} + \cdots \tag{6.14c}$$

As a guide, the cross-section should be divided into a total of roughly 10–100 segments along each side length. Then, sufficiently accurate results are usually obtained.

Spreadsheets – or generally matrix-based programming languages – are particularly convenient for this type of solution. Each subarea ΔA_i is assigned a cell containing a specific parameter. These parameters are coordinates y_i and z_i, strain values ε_i, stress values σ_i, and bending moment increments $\sigma_i z_i$ and $\sigma_i y_i$. Summing over the cells yields the axial force N or bending moments M_y and M_z. It is reasonable to ensure that cross-section and spreadsheet area are geometrically similar. In doing so, the cross-section shows its shape in the sheet (cf. Figs. 6.8, 6.10–6.12).

6.4.2 Optimization Problem

Straightforward solvers such as gradient or Newton methods [17, 18] are adequate for solving the optimization problem. Alternatively, evolutionary approaches can also be applied. No particular requirements arise from the problem itself as long as only a moderate number of design variables x_i exists. Difficulty increases with the number of variables. Both zero vectors and boundary values of constraints are inappropriate as starting points. Sound initial design variable vectors can be obtained by preliminary estimation, for example by determining the linear elastic solution of the initial cross-section. If the optimizer converges toward a state where equilibrium is not achieved, the solution is invalid. It should be noted that, since the problem is generally non-convex, finding the global minimum is not guaranteed. This issue can be overcome by varying the initial vector of optimization variables in order to obtain multiple solutions and take the one yielding the lowest objective function values as the optimum.

6.5 Parameterization

In order to avoid treating each cross-section as an individual case – which would be highly ineffective in daily engineering practice, where cross-section types and loading scenarios repeat – parameterization is recommended. Cross-sections are thereby addressed depending on their basic type. In doing so, a cross-section type refers to a fundamental shape, a principle distribution of materials (e.g. reinforcement locations) that can be described relative to the basic cross-sectional dimensions, and possible characteristic values of material properties (e.g. strength, Young's modulus). Parameterization should be applied consistently in terms of geometry, material, and discretization. In the case of applying numerical solution approaches, the discretization into the subareas A_i should also be parameterized such that a consistent division results independent of the actual cross-sectional quantities. It is reasonable to relate the widths (Δb) and heights (Δh) of the subareas ΔA to the actual cross-sectional dimensions, namely $\Delta b = b/n_b$ and $\Delta h = h/n_h$. Moreover, a distinction must be made between plane and spatial problems, as will be discussed in the following.

6.5.1 Plane Case

For cross-sections in plane problems, both geometric and mechanical axis symmetry exist. This must also apply to the design variables, hence, the symmetry may not be affected by the optimization. If the symmetry axis coincides with the z-axis, the cross-sectional shape must also be symmetrical to this axis. Consequently, the same applies to the strain plane. As a result, only axial forces and bending moments perpendicular to the mirror plane may exist. All quantities must be a function of the z-coordinate alone and not depend on y-values. Typical examples are cross-sections with rectangular, circular, or box shape.

Figure 6.4 shows a simple rectangular RC cross-section. It exhibits a height h, a width b, and the location of the reinforcement (d_h) relative to the z-direction.

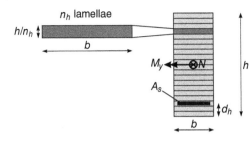

Figure 6.4 Parameterized rectangular cross-section under plane loadings M_y and N discretized by n_h lamellae over its height h.

The reinforcement A_s is represented as a line, which implies that it is invariant in the y-direction. The applied material parameters are f_c, f_y, and f_t, that is, the concrete compressive strength, as well as the yield stress and tensile strength of the reinforcement. The discretization is provided by n_h horizontal lamellae in the z-direction, which form the subareas $\Delta A = \frac{h}{n_h} b$.

A similar concept can be developed for the T-section in Figure 6.5, specifically for the part shown on the right when also the material distribution is symmetrical. Here, additional geometry and reinforcement variables are added. However, the fundamental axis symmetry in geometry, reinforcement, and actions is maintained.

6.5.2 Spatial Case

Spatiality arises from either lack of geometric symmetry, spatial loading involving two bending moments (M_y, M_z), or lack of a material distribution symmetry. In this respect, the presence of one of the criteria is sufficient. However, they may also appear combined. In any case, this results in the strain plane being no longer constant perpendicular to a principal axis direction. Instead, it must be described as a function of y- and z-coordinates, which implies that three equilibrium conditions

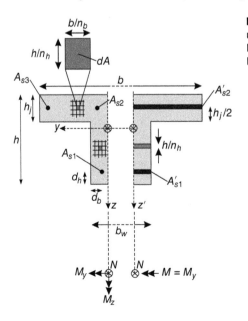

Figure 6.5 Parameterized T-section under spatial (left) or plane (right) loadings discretized into fibers (left) or lamellae (right).

have to be taken into account. For the parameterization, this entails some modifications. First, the reinforcement A_s must now be specified in terms of reinforcement points, meaning that the location in y-direction must be included as a variable. Second, the discretization includes fibers that divide the cross-section along the y- and z-directions into subareas $\Delta A = \frac{h}{n_h} \cdot \frac{b}{n_b}$, where n_h and n_b are the numbers of fibers over the height and width, respectively. All other parameters are to be defined according to the plane case.

Figure 6.5 left shows the parameterization concept for a T-shaped cross-section. The cross-section is loaded spatially and exhibits three fixed reinforcement points A_{s1}, A_{s2}, and A_{s3}, as well as a discretization in fibers subdivided in y- and z-direction. The lamellae and fibrous divisions can freely vary in size between the individual cross-sectional parts, as long as sufficiently fine subdivision is provided.

It is often easier to consider the reinforcement, which is smaller in area than the area made from the main material, separately and simplistically idealize the full cross-section in lamellae or fibers, respectively, of the main material. This usual approach of the gross cross-section should then be compensated again according to Eq. (2.11) when determining the reinforcement resistance.

6.5.3 Parameterization with Intentional Steering

For practical reasons, it is often necessary to limit the unrestricted formation of a cross-section. For example, symmetries ought to be generated, the outer shape is to be preserved, or a relative reinforcement distribution must be accounted for. A typical parameterization with intentional steering is the latter example, namely pre-specifying the relative reinforcement distribution within the cross-section. Here, a total amount $A_{s,tot}$ is computed, where the individual quantities A_{si} are characterized by a defined proportion $\alpha_i \in [0,1]$ thereof:

$$A_{s,tot} = \sum_i \alpha_i A_{si} \qquad (6.15)$$

For example, a requirement in practice could be that the upper reinforcement A_{s1} is twice as large as the lower one A_{s2}. This yields the weighting variables $\alpha_1 = 1/3$ and $\alpha_2 = 2/3$. Apart from the convenient simplification of the reinforcement arrangement, this predefined relation offers the advantage of leaving only one reinforcement quantity to be determined as design variable x_1, since all others can be expressed with respect to it. Moreover, convergence behavior of the optimization is significantly improved compared to a free adjustment of the reinforcement values, because the number of design variables is reduced.

The steered parameterization is illustrated exemplarily in Figure 6.6 by the example of a rectangular cross-section. It is geometrically characterized by its height h, width b, and the four possible corner reinforcement points, whose locations are specified by d_b and d_h. The cross-section is spatially loaded. The objective here is to calculate the reinforcement quantities at given relations for arbitrary loadings. The total amount of the reinforcement is determined as follows:

$$A_{s,tot} = A_{s1} + A_{s2} + A_{s3} + A_{s4} = (\alpha_1 + \alpha_2 + \alpha_3 + \alpha_4) A_{s,tot} \qquad (6.16)$$

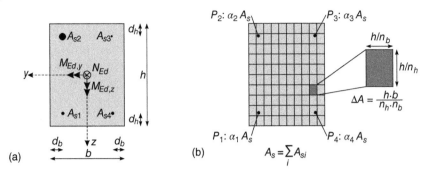

Reinforcement point	y-Coordinate	z-Coordinate	$A_{si}/A_{s,\text{tot}}$
P_1	$b/2 - d_b$	$h/2 - d_h$	α_1
P_2	$b/2 - d_b$	$-h/2 + d_h$	α_2
P_3	$-b/2 + d_b$	$-h/2 + d_h$	α_3
P_4	$-b/2 + d_b$	$h/2 - d_h$	α_4

(c)

Figure 6.6 Parameterized rectangular cross section under spatial loading with four individual reinforcement points P_1 to P_4: (a) denotations and relative geometry, (b) relative reinforcement amounts and discretization in $n_b \times n_h$ fibers, (c) related locations and amounts of the reinforcement points.

where the weighting factors α_i are predefined by the user. All fractions α_i must add to 1, which equals 100 % of $A_{s,\text{tot}}$. The (gross) concrete area is discretized into rectangular fibers, each divided proportionally to the height (n_h) and width (n_b):

$$\Delta A = \frac{h \cdot b}{n_h \cdot n_b} \tag{6.17}$$

With this one parameterized model of a rectangular cross-section, all possible interaction diagrams – whether for plane or spatial loading, independent of the pre-defined reinforcement distribution and for all concrete strength classes – are solved at once and can be used without the need of manual interpolation. This shows the significant advantage of the universally valid parameterization for the daily engineering practice.

6.6 Examples

6.6.1 Equilibrium Iteration

Example 6.1 (Strain plane of an unsymmetric RC section). A complex RC section shown in Figure 6.7 is to be designed in ULS. The border of the cross-section consists of a stepped outer shuttering shape. The reinforcement distribution and

Figure 6.7 Unsymmetric cross-section with given loading $N_{Ed} = -50$ kN, $M_{Ed} = 280$ kNm, reinforcements and material parameters of concrete and steel.

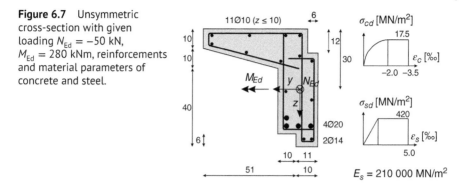

quantities are given, as well as the spatial design actions $N_{Ed} = -50$ kN and $M_{Ed} = 280$ kNm. Longitudinal reinforcement is provided at the stirrup corners, as intermediate bars and in the form of main tensile reinforcement at the lower border (4Ø20 + 2Ø14). For the given boundary conditions, the stress and strain distribution in ULS is sought [12, 19, 20].

The concrete exhibits a design compressive strength of $f_{cd} = 17.5$ MPa at a limit strain $\varepsilon_{cu} = -3.5$ ‰, which corresponds approximately to type C30/37 according to Eurocode 2 [2]. The parabola–rectangle diagram from Section 2.4.1 is used as constitutive model to describe the stress–strain relation. Hence, no tensile load capacity is applied. The reinforcement is made of steel, modeled using a bilinear elasto-plastic material law (see Section 2.4.4). The plastic branch is assumed to be horizontal and thus yield stress and ultimate tensile strength are equal, namely $f_{yd} = f_{td} = 420$ MPa. The elongation limit amounts to $\varepsilon_{tu} = 5$ ‰ and the Young's modulus is assumed to be $E_s = 210\,000$ MPa.

The concrete cross-section is uniformly discretized into 1 cm × 1 cm fibers, resulting in a total of 1728 units (Figure 6.8). For simplicity, the inclination of the upper cantilever is modeled in a stepped manner. Constant concrete stress is assumed in

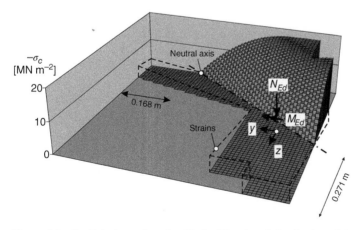

Figure 6.8 Spatial plane of strains (dashed lines) and distribution of the concrete compressive stresses.

each partial area ΔA_i. The rebars are idealized as points. The optimization problem is posed as a minimization of the squared errors according to Eq. (6.3) using the exponent $p = 2$.

The problem is solved using a standard gradient method and the computing time takes only a few seconds for standard office computers. The initial design variable vector for the optimization can be chosen arbitrarily. The optimal solution of the strain plane equation in the ULS finally yields

$$\varepsilon(y, z) = 0.424 + 9.340z + 6.974y \tag{6.18}$$

Figure 6.8 illustrates the resulting strain and stress distributions of the concrete area.

Example 6.2 (Footing with gapping joint). A rectangular footing of the size 8 m × 6 m (Figure 6.9) is loaded by a vertical force $V = -3$ MN and two bending moments $M_x = -3.9$ MNm and $M_y = 5.4$ MNm [13]. For the sake of convention, the coordinate system of the cross-section is denoted here by x and y, whereas the compressive force V acts in the z-direction. For uniform bedding with a constant modulus of subgrade reaction (E_{bed}), the contact pressure distribution on the soil is to be computed. For this purpose, the stress is assumed proportional to the pressure u in the soil (trapezoidal stress method [21]). Due to the double eccentricity $e_x = V/M_y = 1.8$ m and $e_y = V/M_x = 1.3$ m exceeding the first kernel width, gapping occurs (Figure 6.9). Thus, in addition, the load V can also be represented by the force $R = V$, which is shifted from the centroid by the eccentricities.

For the numerical integration, the footing cross-section is divided into 20 × 20 subareas, each of the dimension $\Delta A_i = 0.40$ m × 0.30 m = 0.12 m². This results in a total of 400 fibers. In each fiber, the stress is assumed to be constant.

The optimization problem is formulated as a minimization of the squared errors ($p = 2$) according to Eq. (6.3). The initial design variable vector of b_1, b_2, and b_3 can be chosen arbitrarily apart from the zero vector. A simple gradient method is applied

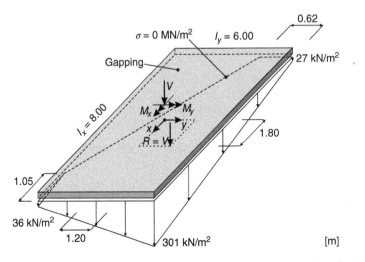

Figure 6.9 Rectangular footing (8 m × 6 m) under eccentric loading $R = V$ with distribution of soil stresses and gapping joint.

Soil stresses [kN/m²]

	1	2	3	4	5	6	7	8	9	10	11	12	13	14	15	16	17	18	19	20
1	-36	-22	-8	0	0	0	0	0	0	0	0	0	0	0	0	0	0	0	0	0
2	-49	-35	-22	-6	0	0	0	0	0	0	0	0	0	0	0	0	0	0	0	0
3	-62	-49	-35	-21	-6	0	0	0	0	0	0	0	0	0	0	0	0	0	0	0
4	-76	-62	-48	-35	-21	-7	0	0	0	0	0	0	0	0	0	0	0	0	0	0
5	-89	-75	-61	-48	-34	-20	-7	0	0	0	0	0	0	0	0	0	0	0	0	0
6	-102	-89	-75	-61	-47	-34	-20	-6	0	0	0	0	0	0	0	0	0	0	0	0
7	-115	-102	-88	-74	-61	-47	-33	-19	-6	0	0	0	0	0	0	0	0	0	0	0
8	-129	-115	-101	-88	-74	-60	-46	-33	-19	-5	0	0	0	0	0	0	0	0	0	0
9	-142	-128	-115	-101	-87	-73	-60	-46	-32	-19	-5	0	0	0	0	0	0	0	0	0
10	-155	-141	-128	-114	-100	-87	-73	-59	-46	-32	-16	0	0	0	0	0	0	0	0	0
11	-168	-155	-141	-127	-114	-100	-86	-73	-59	-45	-21	-18	0	0	0	0	0	0	0	0
12	-182	-168	-154	-141	-127	-113	-99	-86	-72	-58	-45	-31	-11	0	0	0	0	0	0	0
13	-195	-181	-168	-154	-140	-126	-113	-99	-85	-72	-58	-44	-31	-11	-3	0	0	0	0	0
14	-208	-194	-181	-167	-153	-140	-126	-112	-99	-85	-71	-57	-44	-30	-16	-3	0	0	0	0
15	-221	-208	-194	-180	-167	-153	-139	-126	-112	-98	-84	-71	-57	-43	-30	-16	-2	0	0	0
16	-235	-221	-207	-194	-180	-166	-153	-139	-125	-111	-98	-84	-70	-57	-43	-29	-16	-2	0	0
17	-248	-234	-221	-207	-193	-179	-166	-152	-138	-126	-111	-97	-84	-70	-56	-42	-29	-15	-1	0
18	-261	-248	-234	-220	-206	-193	-179	-165	-152	-138	-124	-111	-97	-83	-69	-56	-42	-28	-15	-1
19	-274	-261	-247	-233	-220	-206	-192	-179	-165	-151	-137	-124	-110	-98	-83	-69	-55	-42	-28	-14
20	-288	-274	-260	-247	-233	-219	-206	-192	-178	-164	-151	-137	-123	-110	-96	-82	-68	-55	-41	-27

Figure 6.10 Numerical results of block wise constant soil stresses in a spreadsheet.

for solving and yields the deformation plane $u(x, y)$:

$$u(x,y) = b_1 + b_2 x + b_3 y \tag{6.19}$$

where the deformations [m] multiplied by the modulus of the subgrade reaction [kN/m³] give the soil stresses σ [kN/m²]. It should be noted that the constant modulus of subgrade reaction E_{bed} is a value of no relevance for the stress distribution. It can be freely chosen. Stresses are computed independently of E_{bed}, only the extent of the deformations u change proportionally.

Figure 6.10 shows the compressive stress distribution calculated for each subarea. The calculation is implemented in a spreadsheet software in such a way that the foundation area and the table area for stress calculation are of the same shape. The division into 20×20 fibers is distinctly visible with the x-direction pointing to the left and the y-direction pointing downward. The vertical load is offset to the lower left from the center, exactly coinciding with the center of gravity of the soil stress distribution.

Edge and corner stresses can be calculated directly from the deformation plane by inserting the corresponding x and y values, and E_{bed}. Alternatively, they can be interpolated approximately from the values of the table cells. Figure 6.9 shows the linear stress distribution determined in this way. The maximum compressive stress equals 301 kN/m² and is located at the front right corner, whereby a subgrade reaction modulus of $E_{bed} = 10\,000$ kN/m³ is assumed. The large eccentricity results in wide gapping regions on the opposite side, which do not contribute to the load transfer. For this reason, optimizing the foundation shape is reasonable to enhance efficiency, as will be done in Example 6.6.

Example 6.3 (Parameterized T-section). For arbitrary T-sections under spatial loading, a general parameterized solution is set up [15]. It follows Figure 6.5 (left) and may now be used in this specific example to determine the strain and stress distributions at ULS for a defined cross-section and given design actions $N_{Ed} = -100$ kN, $M_{Ed,y} = 600$ kN m, and $M_{Ed,z} = -150$ kN m.

The corresponding parameterization is depicted in Figure 6.12. The flange and the web consist of 10×100 ($\Delta A_f = h_f/10 \times b/100$) and 20×40 ($\Delta A_w = h_w/40 \times b_w/20$)

subareas, respectively. Six reinforcement points A_{s1} to A_{s6} are specified. A_{s1} and A_{s2} have 3Ø25 (14.7 cm²) each, whereas A_{s3} to A_{s6} correspond to 2Ø10 (1.57 cm²) each. The rebar point position is defined by the horizontal and vertical distances from the cross-sectional edge to their centers of gravity, namely $d_h = d_v = 0.06$ m. For the sake of simplicity, the flange reinforcement is grouped together in the center of each section. The center of gravity of the gross concrete cross-section is located at 0.223 m from the top edge of the flange in the z-direction.

The employed concrete corresponds to the strength class C40/50 according to Eurocode 2 [2], exhibiting a design compressive strength of $f_{cd} = 22.7$ MPa and a limit compressive strain $\varepsilon_{cu} = -3.5$‰. The parabola–rectangle diagram (Section 2.4.1) is applied as the constitutive model to describe the stress–strain relation. In doing so, a tensile load capacity is neglected. The reinforcing steel is of type B500 and the bilinear constitutive diagram with an increasing plastic branch is used. The assumed yield stress and ultimate strength are $f_{yd} = 435$ MPa and $f_{td} = 456$ MPa at the elongation values $\varepsilon_{yd} = 2.174$‰ and $\varepsilon_{td} = 25$‰, respectively. The modulus of elasticity is assumed to be $E_s = 200\ 000$ MPa.

The distribution of strains and stresses is to be determined. For this purpose, the optimization problem is formulated as minimization of the error squares with $p = 2$ according to Eq. (6.3). The resulting strain plane as a solution of the optimization problem is computed numerically using a standard gradient method. This then yields

$$\varepsilon(y, z) = 0.141 + 6.728z + 4.389y \tag{6.20}$$

where the coordinates y and z are given in [m] and the respective elongation is expressed in [‰].

Figure 6.11 and 6.12 show the distribution of the concrete stresses in ULS as a spatial graph and as cell values of a spreadsheet, respectively. The highest concrete stresses are found at the upper right corner for an edge strain of −2.7‰. Hence,

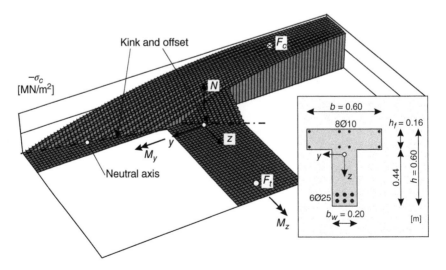

Figure 6.11 T-section with geometry parameters, reinforcements, and concrete stress distribution illustrated by columns over a fiber pattern.

Concrete stresses [MN/m²]

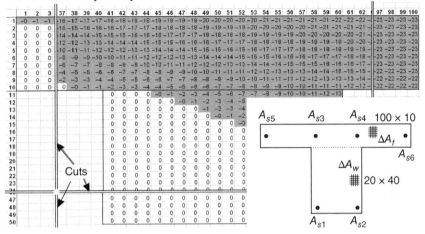

Figure 6.12 Parameterized discretization of the T-section and calculated concrete stresses within the corresponding fiber pattern in a spreadsheet.

the concrete is not fully utilized ($> -3.5‰$), although being within its plastic capacity ($< -2‰$). The resulting moment $M_{Ed} = \sqrt{M_{Ed,y}^2 + M_{Ed,z}^2}$, with its vector pointing inclined from bottom right to top left, forms the compressive stresses as internal response quantities. The cell values yield the anticipated pattern of the parabola–rectangle diagram. The resulting tensile force of the reinforcement is located roughly at the center of the lower reinforcements $A_{s1} + A_{s2}$, with a slight offset in the positive y-direction.

The representation of the compressive stresses as columns provides an equidistant division of web and flange fibers. Note that this is solely due to the graphical representation, which does not reflect the actual fiber dimensions. Based on the parameterized division, the fibers have widths in y-direction of $b_w/20 = 0.01$ m for the web and $b/100 = 0.006$ m for the flange. The same applies to the heights with $(h - h_f)/40 = 0.011$ m for the web and $h_f/10 = 0.016$ m for the flange. Both differ in cell sizes. The resulting shift and kink of the neutral axis in Figure 6.11 and 6.12 is thus owed exclusively to the adopted graphical visualization.

6.6.2 Sectional Optimization

Example 6.4 (Parameterized uniaxial bending). A fixed, rectangular ($b/h/d_h = 0.25/0.50/0.05$ [m]) RC cross-section is subjected to a bending moment M_{Ed}. It is made of concrete type C20/25 [2] and exhibits different reinforcement amounts at its lower ($A_{s,bot}$) and upper ($A_{s,top}$) side. M_{Ed} should have different values between 100 and 400 kNm. The minimum required total reinforcement $A_s = A_{s,bot} + A_{s,top} \rightarrow$ min and its distribution between the top and bottom layers is to be determined. This should be done for the different values of M_{Ed}. In doing so, the well-known transition from initially one-sided bending reinforcement (tension zone decisive, compression zone of the concrete sufficiently load-bearing) to the installation of additional compression reinforcement (two-sided reinforcement for

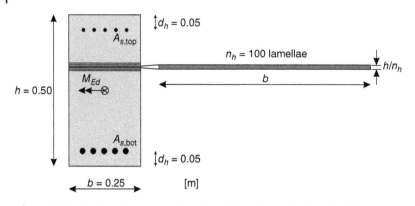

Figure 6.13 Rectangular cross-section in axial bending with discretization.

additional strengthening of the compression zone while maintaining a favorable strain state) will be shown (Figure 6.13).

The compressive behavior of the concrete is approximated with the parabola–rectangle diagram (Section 2.4.1). Its design strength equals $f_{cd} = 11.3$ MPa and the ultimate elongation is assumed to be $\varepsilon_{cu} = -3.5\%_0$. Reinforcing steel B500 with elasto-plastic material law and hardening behavior is employed, resulting in $f_{yd}/f_{td} = 435/456$ MN/m^2 at a yield and ultimate strain $\varepsilon_{yd}/\varepsilon_{td} = 2.174/25\%_0$. The modulus of elasticity is $E_s = 200\ 000$ MPa.

For numerical integration, the concrete cross-section is discretized over its height in $n_h = 100$ lamellae, similarly to Figure 6.4. The resulting partial areas $\Delta A_i = h/n_h b$ exhibit uniform concrete stresses each. The reinforcements are idealized as lines having their center of gravity located at $d_h = 0.05$ m from the upper and lower edges.

A plane problem exists where the minimum amount of reinforcement is sought. Thus, the design variable vector contains, on the one hand, the parameters for constructing the strain plane (b_1, b_2) and, on the other hand, the upper and lower reinforcement quantities $(A_{s,top}, A_{s,bot})$:

$$\mathbf{x}^\top = \begin{bmatrix} b_1 & b_2 & A_{s,top} & A_{s,bot} \end{bmatrix} \tag{6.21}$$

The objective is to minimize the total amount of reinforcement:

$$f(\mathbf{x}) = A_{s,top} + A_{s,bot} \rightarrow \min \tag{6.22}$$

In order to ensure equilibrium of forces, an equality constraint according to Eq. (6.5) is introduced involving the normal force $N_{Ed} = 0$ and the bending moment M_{Ed}. Any concrete stress components can be subtracted from the steel reinforcement areas, since they are already taken into account over the full gross cross-section (cf. Eq. (2.11)). For the sake of simplicity, however, this example is calculated without this subtraction. Thus, gross sectional concrete values are assumed.

Moreover, inequality constraints limit the concrete strains at the upper edge (g_1), on the one hand, and, on the other hand, the steel strains in the lower reinforcement layer (g_2):

$$g_1(\mathbf{x}) = \varepsilon_{cu} - \varepsilon_{c,top} \leq 0 \tag{6.23a}$$

$$g_2(\mathbf{x}) = \varepsilon_{s,bot} - \varepsilon_{tu} \leq 0 \tag{6.23b}$$

Table 6.1 Bending moment and required amounts of bottom and top reinforcement for a rectangular RC cross-section.

M_{Ed} [kN m]	req. $A_{s,bot}$ [cm²]	req. $A_{s,top}$ [cm²]	$\varepsilon_c/\varepsilon_s$ [‰]
100	5.6	0	
200	13.3	0	var.
213	14.6	0	
220	15.0	0.4	
300	19.6	5.0	const. $= \varepsilon_{c,top}/\varepsilon_{s,bot} = \varepsilon_{cu}/\varepsilon_{yd}$
400	25.4	10.7	

Two additional inequality constraints further enforce only nonzero values for the reinforcement quantities:

$$g_3(\mathbf{x}) = -A_{s,top} \leq 0 \tag{6.24a}$$

$$g_4(\mathbf{x}) = -A_{s,bot} \leq 0 \tag{6.24b}$$

Table 6.1 shows six magnitudes of the moment with associated amounts of reinforcement for the upper and lower layer computed by solving the optimization problem via a standard gradient method. First, only tensile (lower) bending reinforcement is calculated, which increases slightly nonlinear with the size of M_{Ed}. At approximately $M_{Ed} = 213$ kN m, a strain distribution $\varepsilon_{c,top}/\varepsilon_{s,bot}$ of $\varepsilon_{cu}/\varepsilon_{yd} = -3.5/2.174$‰ is obtained. Subsequently, this strain state is maintained, and almost equal upper and lower reinforcement is added to cover the increase in bending moment. This corresponds exactly to the expected, cost-effective approach.

Example 6.5 (Shape design of a RC I-section). The cross-sectional design of a RC beam with I-section loaded by a positive bending moment $M_{Ed} = 1000$ kN m is sought. Its height h should amount to 1.00 m, the flange b_f and web width b_w to 0.40 m and 0.06 m, respectively, and the distance d_h to 0.05 m. Both the concrete flange height h_f on the compressed upper side and the required reinforcing steel amount A_s on the lower tensile side are to be optimized. In doing so, different "costs" for the two materials are to be assumed. Three scenarios will be examined:

- Steel is 10 times more "expensive" than concrete (reality, $a_s/a_c = 10$)
- Steel and concrete are equally "expensive" (parity, $a_s = a_c = 1$)
- Concrete is 10 times more "expensive" than steel ("inverted reality", $a_s/a_c = 0.1$)

These costs are applied as multipliers to the areas of steel or concrete. Cross-sectional designs minimizing the total amount of concrete and steel weighted by a_c and a_s, respectively, are to be determined. It should be noted that typical price ratios are in the order of magnitude beyond $a_s/a_c \geq 10$, depending on the types of concrete and reinforcing steel. The first scenario with significantly more "expensive" steel is thus a (currently) realistic one.

The symmetric I-section with its respective dimensions is shown in Figure 6.14. The reinforcement cross-sectional area A_s and the flange height h_f represent the design variables that have to be determined. The employed material is concrete type C30/37 according to [2]. Its stress–strain behavior is approximated using the parabola–rectangle diagram (Section 2.4.1). The ultimate elongation is assumed to be $\varepsilon_{cu} = -3.5‰$ and its design compressive strength equals $f_{cd} = 17$ MPa. No tensile bearing capacity is taken into account. The reinforcing steel is modeled bilinearly with increasing plastic branch (cf. Section 2.4.4). The type corresponds to B500 with stress limits $f_{yd}/f_{td} = 435/456$ MPa and corresponding strain values $\varepsilon_{yd}/\varepsilon_{td} = 2.174/25‰$. The Young's modulus equals $E_s = 200\ 000$ MPa.

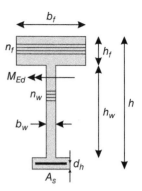

Figure 6.14 I-shaped RC section under bending, discretized into lamellae, with A_s and h_f being the design variables.

The cross-section is divided into a total of 100 lamellae, 50 over the flange height h_f, and 50 over the remaining web height $h_w = h - h_f$. As h_f alters, the lamella thicknesses vary accordingly and thus the cross-sectional areas become $\Delta A_{cf} = h_f/50 \cdot b_f$ and $\Delta A_{cw} = h_w/50 \cdot b_w$. The reinforcement A_s is idealized as a line with its center of gravity located at $d_h = 0.05$ m from the lower edge.

The optimization variable vector reads

$$\mathbf{x}^\mathsf{T} = \begin{bmatrix} b_1 & b_2 & A_s & h_f \end{bmatrix} \tag{6.25}$$

where b_1 and b_2 describe the strain plane according to Eq. (2.4). The objective function is formulated according to Eq. (6.13) as minimizing the concrete (flange) and steel (A_s) cross-sectional areas weighted by their corresponding cost multiplier:

$$f(\mathbf{x}) = \alpha_s A_s + \alpha_c h_f b_f \rightarrow \min \tag{6.26}$$

Equality constraints applied to the normal force $N_{Ed} = 0$ and the bending moment M_{Ed} ensure equilibrium between internal and external forces of the solution:

$$h_1 = N_{Ed} \overset{!}{=} \sum_i^{n_f} \sigma_{ci}\Delta A_{cf} + \sum_i^{n_w} \sigma_{ci}\Delta A_{cw} + \left(\sigma_{s,\text{bot}} - \sigma_{c,\text{bot}}\right) A_s = 0 \tag{6.27a}$$

$$h_2 = M_{Ed} \overset{!}{=} \sum_i^{n_f} \sigma_{ci}z_i\Delta A_{cf} + \sum_i^{n_w} \sigma_{ci}z_i\Delta A_{cw} + \left(\sigma_{s,\text{bot}} - \sigma_{c,\text{bot}}\right) z_{\text{bot}} A_s \tag{6.27b}$$

Any concrete stress components are subtracted from the areas of the steel reinforcement, since they are already taken into account over the total gross cross-section. The integrals also distinguish between concrete components in the flange (i to n_f) and those in the web (i to n_w).

Additionally, a total of four inequality constraints are imposed, which limit the concrete compression at the top edge (g_1), as well as the steel strains in the tensile reinforcement A_s (g_2), and further enforce physically reasonable (positive) values for

Table 6.2 Reinforcement area A_s, flange height h_f, and minimized area $A^* = \alpha_s A_s + \alpha_c b_f h_f$ as functions of weighting factors α_s/α_c.

α_s/α_c [−]	$A_{s,bot}$ [cm²]	h_f [m]	A^* [cm²]
10/1	26.7	0.182	997
1/1	80.5	0.143	653
1/10	260.0	0.129	5419

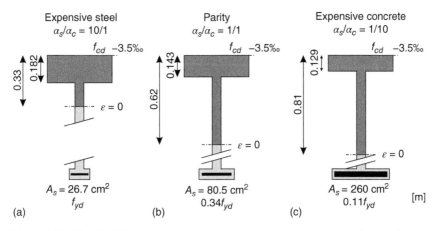

Figure 6.15 Required flange areas of concrete under compression and rebars under tension for (a) "expensive" steel, (b) parity between concrete and steel, (c) "expensive" concrete.

A_s and h_f (g_3, g_4):

$$g_1(\mathbf{x}) = \varepsilon_{cu} - \varepsilon_{c,top} \leq 0 \qquad (6.28a)$$

$$g_2(\mathbf{x}) = \varepsilon_s - \varepsilon_{tu} \leq 0 \qquad (6.28b)$$

$$g_3(\mathbf{x}) = -A_s \leq 0 \qquad (6.28c)$$

$$g_4(\mathbf{x}) = -h_f \leq 0 \qquad (6.28d)$$

Table 6.2 shows the calculated reinforcement steel quantities A_s, the flange height h_f, and the total area A^* of the objective function to be minimized resulting from the applied weighting factor ratios α_s/α_c. The total area is defined as $A^* = \alpha_s A_s + \alpha_c b_f h_f$, that is, a weighted sum of tension and compression flange subareas. Figure 6.15 depicts the determined cross-sections for the three different scenarios. In the case of expensive steel, which corresponds to the current situation of actual costs, a small amount of less than 27 cm² is used, thus meeting its yield strain $\varepsilon_{yd} = 2.174\%$. On the compressed side, the concrete is fully utilized with $\varepsilon_{cu} = -3.5\%$, yielding a flange thickness of roughly 18 cm. This provides a practical solution. Making the steel more expensive (parity and 10/1) results in the steel being utilized far below its yield stress (34 % and 11 % of f_{yd}). Accordingly, about three and almost 10 times as much rebar cross-sectional areas is needed as with the "expensive steel" scenario, respectively. Due to the high price of concrete, the compression zone is extended far into the web, resulting in the smallest possible flange area of 14.3 and 12.9 cm in height, respectively.

Example 6.6 (Shape optimization of a footing). An eccentrically loaded footing according to Example 6.2 exhibits large gapping areas (Figure 6.16). The gapping area constitutes almost 40 % of the overall foundation area and does not contribute to the load transfer. $V = -3$ MN is offset from the column by 1.8 m in the x-direction and by 1.3 m in the y-direction. This yields the bending moments $M_x = -3.9$ MN m and $M_y = 5.4$ MN m. The maximum edge stress is 301 kN/m² (cf. Example 6.2). The footing is obviously shaped unfavorably and shall be optimized with the aim to minimize the overall area. Thus, the optimization relates to minimizing the amount of material required for the footing size. In doing so, no higher soil stresses should develop than before. They are limited to 300 kN/m².

Preferably, the footing center of gravity should be below the load V. If this is the case, the overall area becomes minimal and a uniform stress distribution establishes. However, the design space is constrained by surrounding buildings to the left and at the lower edge (geometric border). Furthermore, the footing area must enclose the column. For this reason, the side length in the y-direction may not be changed ($l_y = 6.00$ m), which further limits the freedom of shaping. In order to move the center of gravity of the foundation closer to the V load and at the same time minimize its total area, three means are provided:

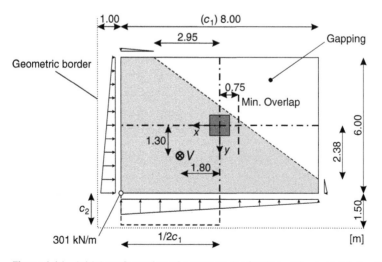
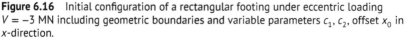

Figure 6.16 Initial configuration of a rectangular footing under eccentric loading $V = -3$ MN including geometric boundaries and variable parameters c_1, c_2, offset x_0 in x-direction.

- The upper longitudinal side c_1 is variable.
- The foundation can be shifted as a whole in x-direction (offset x_0). However, a minimum overlap to the column of 0.75 m must be ensured.
- At the lower end, it is possible to extend the foundation using the additional variable depth c_2, but c_2 must be equal to half of c_1.

The footing cross-section (base area) is discretized into 20×20 subareas (400 fibers ΔA_1) of dimensions $c_1/20$ (variable with c_1) and $6.00 \text{ m}/20 = 0.30 \text{ m}$. The potential footing extension of $1/2 c_1 \cdot c_2$ at the lower edge is discretized into 10×10 subareas (100 fibers ΔA_2) of the dimensions $c_1/10$ (variable with c_1) and $c_2/10$ (variable with c_2). In each subarea ΔA_1 or ΔA_2, the soil stress is considered constant. The foundation is assumed to be a rigid body. The soil stresses are determined via the deformation plane $u(x, y)$ defined by the parameters b_1, b_2 and b_3. These deformations [m] multiplied by the modulus of subgrade reaction [kN/m^2] result in the required contact pressure [kN/m^2]. The design variable vector thus takes the form:

$$\mathbf{x}^{\mathsf{T}} = \begin{bmatrix} b_1 & b_2 & b_3 & c_1 & c_2 & x_0 \end{bmatrix} \tag{6.29}$$

The objective function is formulated following Eq. (6.12) as a minimization of the total footing area A, which is composed of the rectangular portion $A_1 = c_1 \cdot 6.00 \text{ m}$ and its lower extension $A_2 = 1/2 \cdot c_1 c_2$. This yields

$$f(\mathbf{x}) = A = A_1 + A_2 = c_1 \cdot 6 + 1/2 \cdot c_1 c_2 \rightarrow \min \tag{6.30}$$

The vertical load V and the resulting bending moments M_x and M_y caused by the eccentricity yield a total of three equality constraints:

$$h_1 = V \overset{!}{=} \sum_{i=1}^{400} \sigma_i \Delta A_1 + \sum_{j=1}^{100} \sigma_j \Delta A_2 = -3 \text{ MN} \tag{6.31a}$$

$$h_2 = M_x \overset{!}{=} \sum_{i=1}^{400} \sigma_i y_i \Delta A_1 + \sum_{j=1}^{100} \sigma_j y_j \Delta A_2 = -3.9 \text{ MN m} \tag{6.31b}$$

$$h_3 = M_y \overset{!}{=} -\sum_{i=1}^{400} \sigma_i x_i \Delta A_1 - \sum_{j=1}^{100} \sigma_j x_j \Delta A_2 = 5.4 \text{ MN m} \tag{6.31c}$$

The partial areas ΔA_1 and ΔA_2 remain constant in each optimization step.

The limitation of the soil pressure to maximum compressive stresses of 300 kN m^{-2} is ensured by an inequality constraint (g_1). Only the five points at the outward-facing vertices of the foundation are to be checked. Internal values as well as the inward recessed corner in the center of the lower edge are of no relevance. Further inequality constraints ensure sufficient overhang to the right of the support (g_2), compliance with the left boundary (g_3), and that c_1 and c_2 can only take positive values (g_4, g_5). One last constraint ensures that the extension c_2 does not exceed the design space in the y-direction (g_6). The resulting set of inequality constraints thus follows as:

$$g_1(\mathbf{x}) = -300 - \sigma_{\min}(\mathbf{x}) \leq 0 \tag{6.32a}$$

$$g_2(\mathbf{x}) = 0.75 - x_0 - 1/2 c_1 \leq 0 \tag{6.32b}$$

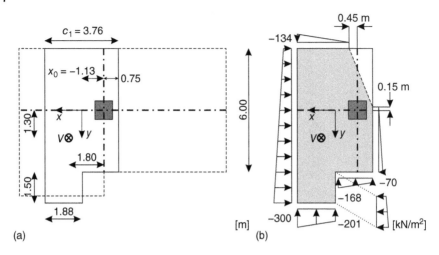

Figure 6.17 (a) Optimized geometry of the footing, (b) distribution of soil stresses.

$$g_3(\mathbf{x}) = 1/2c_1 - x_0 - 5.00 \leq 0 \qquad\qquad (6.32c)$$

$$g_4(\mathbf{x}) = -c_1 \leq 0 \qquad\qquad (6.32d)$$

$$g_5(\mathbf{x}) = -c_2 \leq 0 \qquad\qquad (6.32e)$$

$$g_6(\mathbf{x}) = c_2 - 1.50 \leq 0 \qquad\qquad (6.32f)$$

The starting point of the optimization (superscript 0) is the initial situation shown in Figure 6.16. The stress distribution resulting from the deformation plane $u(x,y)$ described by b_1^0, b_2^0, and b_3^0 is calculated using a subgrade reaction modulus of $E_{\text{bed}} = 10\,000$ kN/m^3. In addition, the initial design variables describing the geometry are assumed to be $c_1^0 = 8.00$ m, $c_2^0 = 0$ m, and $x_0^0 = 0$ m.

Figure 6.17a shows the optimization result after 21 iterations. The upper length is reduced by approximately 53 % compared to the initial value to $c_1^* = 3.76$ m. The lower cross-sectional extension is accordingly half as long (1.88 m) and is extended to the maximum possible boundary ($c_2^* = 1.50$ m). An offset $x_0^* = 1.13$ m of the total cross section shifts the foundation further to the load and is restrained by the minimum overhang of 0.75 m to be maintained. This results in a total area of $A^* = 25.4$ m^2, which reduces the initial size $A^0 = 8.00$ m $\times$ 6.00 m $= 48$ m^2 almost by half. The evaluation of contact pressure in Figure 6.17b shows that the stress requirement of no more than 300 kN/m^2 is precisely met at the lower left corner. This is equivalent to the inherent ULS design of the foundation, which is automatically provided via the constraint during the optimization of the cross-section. As a result, a much smaller gapping area than in the initial design exists at the upper right edge. With a size of 1.71 m^2, it now covers only about 7 % of the total area.

A generalized reduced gradient method was used as optimization algorithm [13]. Figure 6.18 shows the evolution of the individual optimization variables c_1, c_2, and x_0 as well as the objective function $f(\mathbf{x}) = A$ over the 21 steps of iterative computation in a relative fashion. The values are related to their respective final value (superscript *).

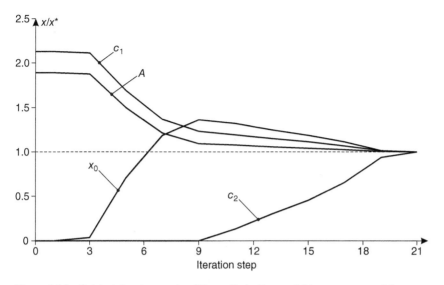

Figure 6.18 Related developments of the optimization variables c_1, c_2, x_0 and the objective function $f(\mathbf{x}) = A$ over the 21 iteration steps of optimization.

They therefore converge toward 1. Due to their trivial initial values, c_2 and x_0 increase from zero. The total area A (objective function) is reduced by about 47 %. A similar result is obtained for the value c_1, where a reduction by 53 % is observed, and from which the major portion of the total area reduction originates. Due to the higher sensitivities, the variables c_1 and x_0 evolve first (iteration step 3), whereas c_2 is subsequently increased to the limit value of 1.5 m starting only from the 9th iteration step.

References

1 Féderation Internationale du Béton (2013). *fib Model Code for Concrete Structures 2010*. Ernst & Sohn. ISBN: 9783433604090.

2 EN 1992-1-1 (2011). *Eurocode 2: Design of concrete structures - Part 1-1: General rules and rules for buildings.*

3 ACI Code 318-19 (2019). *Building Code Requirements for Structural Concrete and Commentary.*

4 Look, K., Oettel, V., Heek, P. et al. (2020). Bemessen mit Stahlfaserbeton. In: *Beton-Kalender 2021* (eds. K. Bergmeister, F. Fingerloos, and J.-D. Wörner), 797–874. Berlin: Ernst & Sohn. ISBN: 978-3-433-03301-2.

5 Look, K., Heek, P., and Mark, P. (2019). Practical calculation, design and optimisation of steel fibre reinforced concrete structures. *Beton- und Stahlbetonbau* 114 (5): 296–306. https://doi.org/10.1002/best.201800097.

6 Gödde, L., Heek, P., Mark, P., and Strack, M. (2016). Stahlfaserbeton. In: *Stahlbetonbau* (eds. J. Hegger and P. Mark), C.1–C.64. Berlin: Beuth-Verlag. ISBN: 978-3-410-26443-9.

7 Yang, H., Lam, D., and Gardner, L. (2008). Testing and analysis of concrete-filled elliptical hollow sections. *Engineering Structures* 30 (12): 3771–3781. https://doi.org/10.1016/j.engstruct.2008.07.004.

8 Gho, W.-M. and Liu, D. (2004). Flexural behaviour of high-strength rectangular concrete-filled steel hollow sections. *Journal of Constructional Steel Research* 60 (11): 1681–1696. https://doi.org/10.1016/j.jcsr.2004.03.007.

9 Goris, A. and Schmitz, U.P. (2014). *Bemessungstafeln nach Eurocode 2: Normalbeton, Hochfester Beton, Leichtbeton*. Köln: Bundesanzeiger-Verlag. ISBN: 3846203912.

10 Allgöwer, G. and Avak, R. (1992). Bemessungstafeln nach Eurocode 2 für Rechteck- und Plattenbalkenquerschnitte. *Beton- und Stahlbetonbau* 87 (7): 161–164.

11 Archdeacon, T.J. (1994). *Correlation and Regression Analysis: A Historian's Guide*. University of Wisconsin Press.

12 Mark, P. (2003). Optimisation methods for the bending design of reinforced concrete sections. *Beton- und Stahlbetonbau* 98 (9): 511–519.

13 Mark, P. (2004). Optimisation of foundation areas and calculation of foundation pressures using optimisation methods and spreadsheet analysis. *Bautechnik* 81 (1): 38–43.

14 Mark, P. and Strack, M. (2004). Bending design of arbitrarily shaped steel fibre reinforced concrete sections using optimisation methods. *6th RILEM Symposium Fibre reinforced concrete (BEFIB 2004)*, 965–974. Varenna, Italy.

15 Mark, P. (2006). *Zweiachsig durch Biegung und Querkräfte beanspruchte Stahlbetonträger*. Habilitation. Bochum: Ruhr University Bochum.

16 Krätzig, W.B., Başar, Y., and Wittek, U. (1997). *Tragwerke*. Berlin: Springer-Lehrbuch, Springer. ISBN: 3-540-62440-6.

17 Nocedal, J. and Wright, S.J. (2006). *Numerical Optimization*. New York: Springer. ISBN: 9780387400655.

18 Fletcher, R. (1987). *Practical Methods of Optimization*. Chichester: Wiley. ISBN: 978-0-471-91547-8.

19 Lauer, H. (1983). Schiefe Biegung mit Längskraft bei beliebigen Stahlbetonquerschnitten. *Bauingenieur* 58: 151–157.

20 Konrad, A. (1988). Ermittlung des Dehnungszustands beliebiger Stahlbetonquerschnitte mit dem Newton-Verfahren. *Beton- und Stahlbetonbau* 83 (10): 261–264.

21 Terzaghi, K., Peck, R.B., and Mesri, G. (1996). *Soil Mechanics in Engineering Practice*. New York: Wiley.